SPACE SHUTTLE STORY

SPACE SHUTTLE STORY

ANDREW WILSON

HAMLYN
London·New York·Sydney·Toronto

Photographic acknowledgments
All photographs in this book are from the National Aeronautics and Space Administration, Washington, D.C., with the exception of the following:
Boeing 12; British Aerospace 14, 15; Rockwell International 21, 31, 52; Martin Marietta 33, 35, 120; European Space Agency 81; McDonnell Douglas 89; Orbital Sciences Corporation 117.

Published by Hamlyn Publishing
Bridge House, Twickenham, Middlesex, England

Prepared by Deans International Publishing
52-54 Southwark Street, London SE1 1UA
A division of The Hamlyn Publishing Group Limited
London · New York · Sydney · Toronto

Copyright © The Hamlyn Publishing Group Limited 1986
ISBN 0 600 50325 9

All rights reserved. No part of this publication
may be reproduced, stored in a retrieval system,
or transmitted, in any form or by any means,
electronic, mechanical, photocopying, recording
or otherwise, without the permission of The Hamlyn
Publishing Group Limited.

Printed and bound by Graficromo s.a., Cordoba, Spain

Contents

Foreword 7

The Origins of the Space Shuttle 8

Preparing for Space 18

Leaving the Earth 41

Up and Running 63

Reaching Maturity 82

The Future Is Not Free 98

Shuttle Flight Summary 124

Index 125

This book is dedicated to the memory of those who have given their lives in the conquest of space:

Francis Scobee, Michael Smith, Judith Resnik,
Ellison Onizuka, Ronald McNair, Gregory Jarvis,
Christa McAuliffe
Died 28 January 1986

Georgy Dobrovolsky, Vladislav Volkov,
Viktor Patsayev
Died 30 June 1971

Vladimir Komarov
Died 24 April 1967

Virgil Grissom, Edward White, Roger Chaffee
Died 27 January 1967

'If we die, we want people to accept it. We are in a risky business . . .'
Virgil Grissom, victim of Apollo spacecraft fire

'The future is not free'
President Ronald Reagan, Shuttle crew memorial service, 31 January 1986

Foreword

1986 promised to be a bumper year of space achievements. It began dramatically with the unique flyby of distant planet Uranus by the space probe Voyager 2 on 24 January. At the same time, an international fleet of five spacecraft continued their journey towards Halley's Comet for a March encounter. America's reusable Space Shuttle was aiming for no less than 15 launches during the year with its complement of four orbiters.

The second Shuttle flight in line was number 51L, a rather mundane mission at the end of January notable really only for the presence aboard of Christa McAuliffe, chosen from 11,000 teachers to become the first private citizen in space. The American space agency considered it to be a routine trip with the veteran *Challenger* spacecraft. Twenty-four flights in almost five years had demonstrated the amazing versatility of the reusable spaceplanes. *Challenger* had flown nine times and was something of a favourite with the astronauts. When it lifted off on 28 January powered by its three main engines and the two boosters strapped either side of the propellant-carrying external tank, there was no hint of trouble. But the flight was little more than a minute old, with the craft hurtling skywards at 2000 mph (32000 km/h) some 10 miles (16 km) high over the Atlantic, when tongues of flame began licking around the base of the external tank. The whole vehicle was engulfed in an incandescent fireball a moment later.

The boosters emerged to fly on separately but *Challenger* had disappeared – only a shower of debris remained to disturb the ocean for more than an hour. Of the seven astronauts there was no sign.

This book tells the story of the Space Shuttle, tracing its history from before the Second World War up to the time of the disaster. Twenty-four missions achieved an outstanding series of successes, including the rescues of no less than four ailing satellites. The destruction of *Challenger* has set the American space programme back on its heels but the Shuttle is so important to future plans that it will emerge, like a phoenix, from the ashes.

The Origins of the Space Shuttle

For nearly three decades now, mankind has been sending machines into space – launching satellites into orbit around the Earth, landing scientific instruments on the surfaces of Mars, Venus and the Moon, and flying probes past Mercury, Jupiter, Saturn and Uranus. The probing of the deeper Solar System is still in its initial phase but nearer to home exploitation is taking over from exploration. Communications satellites have been generating revenue since the 1960s (a $100-million system can produce $1000 million during a ten-year orbital lifetime) and weather satellites have become an indispensable part of our lives. Earth-resources satellites now look down upon the planet to help in the exploration for new sources of oil and minerals, improving efficiency in agriculture, urban planning and management of water resources. Satellites now provide precise navigational services, look for pollution, investigate the complex movements of the atmosphere and oceans, and listen for distress beacons.

The revenue and savings arising from these few examples account for more than all the money expended on getting the first man into space and racing to the Moon.

The exploitation and development of near-Earth space is now recognized as the key to establishing a permanent foothold in space over the next few decades. In this endeavour, men and women have a role to play as well as machines. Space stations circling some 300 miles (500 km) high and, eventually, bases on the Moon will begin a gradual human colonization of space. New products and services will emerge from space so that living on Earth will make *less* sense for human beings, not more. And to carry people, equipment, and products to and fro, types of spacecraft as routine in operation as a train or truck will be required. Their forerunner is the Space Shuttle.

The Early Days

Travel beyond the Earth has been described in literature for centuries but it was the early years of this century that saw the practicalities being tackled for the first time. Pioneers such as Konstantin Tsiolkovsky, a Russian schoolteacher, and Hermann Oberth, a Rumanian, laid the theoretical groundwork but it was experimenters such as Robert Goddard, an American physics professor, who began to build real rockets.

Gunpowder rockets had been known for centuries in the East but it was quickly realized that liquid propellants provide more energy, and are more controllable, than their solid counterparts. The Space Shuttle, for reasons discussed later, uses a mixture of the two, as do some others of today's space launchers. Goddard launched the world's first liquid-propellant rocket on 16 March 1926, powered by gasoline and liquid oxygen. The gasoline acted as a fuel and the liquid oxygen as an oxidizer necessary to allow combustion to take place at high temperatures. The resultant exhaust gases left the rocket at great speed backwards, creating forward thrust in accordance with Isaac Newton's Third Law of Motion that 'To every action, there is an equal and opposite reaction'.

Goddard worked alone but rocket work began to attract the attention of larger bodies, particularly the German Army, who saw it as a means of circumventing the Versailles Treaty of the First World War (when such weapons

were unknown and were thus not included in the treaty). The German Society for Space Travel had been formed in 1927 and had met with such success that the Germany Army effectively took over the organization, which included the young brilliant engineer Wernher von Braun, for military purposes. This resulted in the V-2 missiles, which carried warheads of a ton/tonne of high explosive against London and European targets in the closing stages of the Second World War.

Von Braun and most of his important co-workers went to America following the end of hostilities and became an integral part of the U.S. missile and space scene. It emerged in the 1950s that the Soviet Union was developing an Intercontinental Ballistic Missile (ICBM) capable of reaching the U.S.A. with a nuclear warhead. The Americans instituted crash programmes as it was becoming clear that ICBMs would play a major part in world politics. Smaller rockets (Intermediate-Range Ballistic Missiles or IRBMs) were also being built and by the end of the 1950s the Americans had the Redstone (IRBM), Jupiter (IRBM), Thor (IRBM), Atlas (ICBM) and Titan (ICBM) missiles available.

The Space Age began on 4 October 1957 when the Soviets used a modified form of their ICBM to orbit the 184-lb

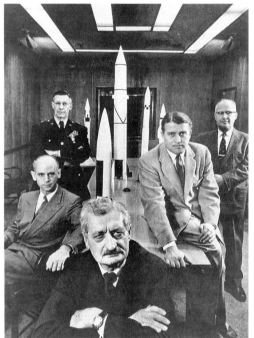

(84-kg) Sputnik 1 satellite. This achievement was then eclipsed by the launch of cosmonaut Yuri Gagarin in his Vostok 1 spacecraft on 12 April 1961. The Americans responded with the Explorer 1 satellite on 1 February 1958 using Von Braun's Redstone, and astronaut John Glenn on 20 February 1962 aboard his Mercury capsule atop a converted Atlas.

Throughout the 1960s and 1970s these kinds of rockets were used for Earth-orbiting satellites and probes to more distant targets – and most were

Above:
Dr. Robert Goddard (1882-1945) launched the first liquid-propellant rocket in 1926. Seen here (second from left) in 1937, Goddard included parachutes on some of his vehicles to allow reuse of components, as with the Shuttle's booster rockets.

Left:
Prof. Hermann Oberth (centre) was a pioneer in theoretical astronautics. Second from right is Wernher von Braun, a major figure in practical rocketry, largely responsible for the V-2, Redstone and large Saturn Moon rockets.

based on the earlier military missiles; a few others, like the small Scout, were not. Even the huge Saturn 5 Moon rockets, standing 365 ft (111 m) tall on the launch pad, as high as St. Paul's Cathedral in London, were based on the same technology, although now liquid hydrogen was being used as a more powerful fuel for the upper stages. The Soviets, likewise, are reported to have built a large vehicle for their manned Moon programme but it failed consistently and was never used in a full mission.

All of these rockets and their satellites or spacecraft had one major feature in common: they were used only once and then discarded after a few minutes in the case of the launch system and at the end of the mission in the case of manned spacecraft. Satellites mostly remained in space or burned up on re-entering the atmosphere. Of course, this meant spaceflight was incredibly expensive – each Saturn 1B flight to take three men to the Skylab space station in 1973/4, for example, cost $120 million at 1972 rates. What was clearly needed was a way of bringing the cost down from about $1000/lb ($2200/kg) of cargo in orbit to a tenth of that value. The obvious method is to reuse as many parts as possible so that new rockets do not have to be built from scratch for every launch. Ideally, the whole spacecraft would be a completely reusable single-stage vehicle with just a few weeks at most between flights. It would provide regular, cheap access to space for payloads and people. In order to land on a conventional runway and to manoeuvre within the atmosphere, it would have wings to produce lift – in other words, it would take off like a rocket, perform its job in space, survive the searing heat of re-entry and then behave like an aircraft on its return to Earth.

Designing a winged spacecraft – never mind the problems of reusability – was by no means a simple undertaking but there was, at least, some experience stretching back, as so often happens, to the German V-2 missile. The normal V-2 had a range of only 150 miles (240 km) and Wernher von Braun's team was interested in increasing this to about 270 miles (430 km) by adding lift-generating wings. The A-4b, as it was called (the V-2 was more properly known as the A-4) would be boosted above the atmosphere so that it could glide on its downward leg to reach its target, enabling the launch site to be as far removed from hostile territory as possible. Such was the state of the war, however, that there was only time to fly two test models. The first, on 8 January 1945, failed soon after launch and the second, 16 days later, broke up, but not before it had become the first winged vehicle to travel faster than the speed of sound.

High Fliers

The first pilot to cross the sound barrier was Charles Yeager flying in the Bell company's rocket-powered X-1 on 14 October 1947. Over the next few years a series of aircraft took men faster and higher – and not without fatalities. The point was approaching in aviation where the dividing line between aircraft and spacecraft becomes blurred. The U.S. National Advisory Committee for Aeronautics decided in 1952 that aerodynamic research should be extended to cover 4 to 10 times the speed of sound and heights of 12 to 50 miles (19 to 80 km). NACA, the U.S. Air Force and the U.S. Navy decided on a joint programme in 1954 and in January 1955 the tag of 'X-15' was given to the project. (The earlier X-series craft had covered the slower, lower regions and the later designations were intended for suborbital and orbital flight.) The three black 50-ft (15.2-m) rocket-powered aircraft that emerged had a solid yet sleek appearance, with small thin wings to provide lift at hypersonic speeds. The X-15 was almost a spacecraft and experience gained with it during the 1950s helped designers to build America's Mercury spacecraft with confidence.

The flights began not on a runway but at altitude below the wing of a modified B-52 bomber flying out of Edwards Air Force Base in California. The X-15 was released and its engine ignited for about 90 seconds to arc the craft above most of the atmosphere. There the sky turned almost to the

Opposite:
The Saturn 5 is still the most powerful rocket ever launched from Cape Canaveral. It last saw service in December 1972, when Apollo 17 blasted off from one of the pads now used by the Shuttle.

The 50-ft (15.2-m) long X-15 rocket aircraft was launched on 199 missions from under the wing of a converted B-52 bomber, reaching a maximum altitude of 67 miles (108 km) in August 1963. Its greatest speed of 4534 mph (7295 km/h) was attained in 1967.

black of space, and the pilot could see the curvature of the Earth and experience a few seconds of weightlessness. The X-15 then re-entered the lower atmosphere, its black skin glowing red with the heat generated by friction with the air, and landed like a glider. Pilot Scott Crossfield made the first powered flight on 17 September 1959 and by the end of 1961 the design speed of six times the speed of sound and a height of 38 miles (61 km) had been reached. No project had ever probed this far and it provided information essential for the design of a winged orbital vehicle.

By the late 1960s, the X-15 was coming to the end of its useful life and budget cuts announced in January 1968 led to the 199th and final flight the following October. Twelve pilots had flown the craft and two of them went on to greater fame: Neil Armstrong (seven flights) became the first man on the Moon and Joe Engle (16 flights) flew the Space Shuttle in its early gliding and orbital tests.

The logical follow-on from the X-15

The X-20 Dynasoar was cancelled in December 1963 as costs soared. The French Hermes and U.S. Air Force vehicle proposals of today are similar.

The lifting body generates lift from the shape of its underside. This M2-F2 model first flew in July 1966 and was rebuilt as the M2-F3 following a crash landing on 10 May 1967.

was a small shuttle-type craft going into a closed path about the Earth and returning from an orbital speed of about 17,500 mph (28,200 km/h) to a conventional landing. The U.S. Air Force decided to go ahead with this Dynasoar project (a contraction of 'dynamic soaring') in 1957 and in November 1959 they selected the Boeing company to build the small, winged craft. Like the X-15, then just beginning its active career, the Dynasoar would be black, but there the resemblance ended. It was short and stubby in appearance with a delta wing and vertical fins, and would be launched by the Titan 1 missile. The aim was eventually to turn it into a military orbital system for surface reconnaissance or inspection of other satellites, but it suffered from many changes of direction during its short life. It gradually became evident that Dynasoar was too ambitious and offered little advantage over the ballistic (i.e. non-lifting) spacecraft then under development for America's first man-in-space project.

The X-24B lifting body flew 36 times during 1973-75, some of the later tests landing on concrete runways at Edwards Air Force Base to simulate a Shuttle return to Earth.

SPACE SHUTTLE FORERUNNERS

X-15: flown 199 times 1959-68 to investigate flight regimes to Mach 6.7.

X-20/Dynasoar: small shuttle craft for orbital/suborbital military missions; never built and cancelled in 1963.

M2, HL-10, X-24A and X-24B: manned lifting bodies to investigate handling of shuttle types at transonic speeds.

ASSET: six launches (1963-65) of small Dynasoar-type models to test thermal protection and aerodynamics.

PRIME: three launches (1966-67) of manoeuvring lifting-body models to near-orbital speeds.

The British Aircraft Corp. MUSTARD (Multi-Unit Space Transport and Recovery Device) was a pioneering shuttle study of the 1960s. Two boosters, possibly manned, peeled off the central orbiter and returned for runway landings and reuse.

The Shuttle Takes Shape

Since 1958, all U.S. civilian space activities have been directed by the National Aeronautics and Space Administration (NASA). Its greatest success came with the culmination of the manned space programmes as Neil Armstrong stepped on to the Moon on 21 July 1969. But NASA was already looking to the future. It saw the age of throwaway rockets coming to an end and the era of the reusable spacecraft and boosters just around the corner. The Space Shuttle concept did not emerge in isolation but as part of the September 1969 report of the Presidential Space Task Group. It recommended a large Space Station, a lunar orbiting base, a lunar surface base and manned missions to Mars before the end of the century – an impressive series of possibilities by any standards. Their achievement required a reusable Shuttle craft to supply the initial 12-man Space Station and a 'Space Tug' to act as a ferry to the Moon and beyond. Not surprisingly, the more exotic elements were rejected on the grounds of cost. NASA thus decided to concentrate solely on the Shuttle, as the Space Station would naturally follow, although they had hoped to have it flying in the 1970s.

First, the agency had to decide what a Shuttle would be required to do before even tentative designs could be formulated. Clearly, it would take satellites into orbit and return them, repair others, house astronauts and experiments for short-duration orbital missions and carry rocket stages for deep-space missions, and it would eventually be used to construct and supply the Space Station. The schemes of 1970 concentrated on a fully reusable Shuttle divided into two parts: the orbiter, carrying cargo and propellant, and a manned booster that would separate about 50 miles (80km) high and fly back to a runway landing under power from aeroengines. The craft were large – one version of the booster

was 257 ft (78 m) long and powered by 12 engines – and the cost was upwards of $10,000 million at 1970 values. The problem of reusability was driving the cost high because designing, building and testing components for repeated use was expensive. An engine operating at thousands of degrees temperature and hundreds of atmospheres pressure was normally thrown away at the end of one mission but to build an engine for up to 100 flights was breaking new ground; the same applied to thermal protection.

Not only was cost a problem but politics and performance intruded. The White House did not want an expensive space programme and NASA had to fight to hang on to the Shuttle at all, having to ally itself with the powerful military lobby. The difficulty here was that military requirements were rather different from what NASA had in mind and a set of compromises had to be reached. The payload bay had to be large: 60 by 15 ft (18.3 by 4.6 m); a cargo of up to 65,000 lb (29,500 kg) had to be carried into low orbit from Cape Canaveral and the design had to allow for a landing up to 1270 miles (2040 km) either side of the re-entry path. An earlier NASA version had aimed at a landing crossrange of only 200 miles (320 km) and a 40,000-lb (18,140-kg) payload – significantly lower capabilities. The large crossrange was required specifically because of polar orbital launches from the Vandenberg Air Force Base so that a Shuttle could return to base after just one circuit of the globe instead of waiting a day for the Earth to rotate and bring the launch site back within reach. This meant that the small wings of NASA's earlier design had to be replaced with much larger versions to generate the necessary lift.

NASA accepted these conditions in January 1971 but the cost of a fully reusable two-craft system was strangling the project. The solution was to concentrate on semireusable components and slash the cost to about $5500 million (1970 values). At first sight, this seems a reasonable compromise but what it did was shift the costs into higher *operational* expenditure, allowing ordinary rockets such as Europe's Ariane to compete successfully against it in the commercial satellite market.

To make the orbiter as small as possible, its propellants would be carried in an external tank, discarded each time, thus adding to the operational cost but keeping the development cost down. On 5 January 1972 it was announced that NASA had received approval from President Nixon for a Shuttle project: 'I have decided today that the United States should proceed at once with the development of an entirely new type of space transportation system.'

At that stage, the final configuration was still not settled. Two boosters either side would ignite at launch before separating at altitude but it was not until 15 March 1972 that NASA announced solid-propellant designs would be chosen in preference to liquid propellants. Again, this allowed a lower development cost (saving an estimated $350 million) but would be more expensive in the long run because they would have to be cleaned and refitted every time – a more complicated process for solids than liquids.

On 26 July 1972 NASA awarded North American Rockwell Corporation a $2600-million contract to act as the main contractor, responsible for building the orbiter portion and integrating it with the external tank and the solid rocket boosters. By May 1973 the detailed design was virtually complete and the appearance of the chunky, delta-winged air/spacecraft familiar to us now was fixed. At this stage, orbital tests flights were expected by 1978 and full operations by 1980 – neither of which came remotely close to being achieved.

Rockwell's 1969/1970 Shuttle concept. The booster has peeled away to land and be refurbished for the next launch.

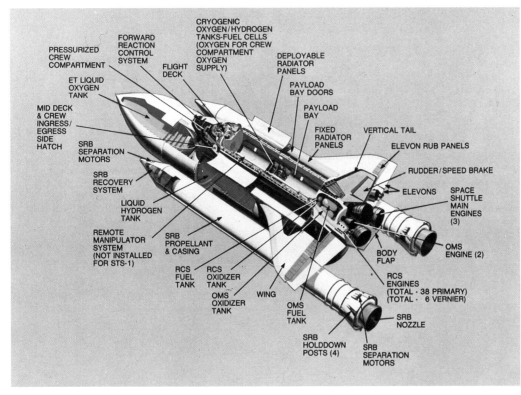

Above:
Wind-tunnel tests were important for predicting the Shuttle's flight characteristics.

Left:
A cutaway view of the complete Shuttle vehicle.

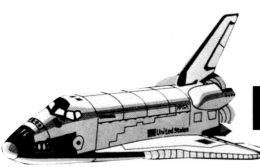

Preparing for Space

Work on the first Space Shuttle, numbered OV-101 (Orbiter Vehicle-101), began in June 1974, although it was not intended to make the first orbital flight. It was essentially a shell of an orbiter to be used for gliding tests and checking compatibility with the complex launch facilities. It would then be returned to Rockwell in California for conversion to a full flying orbiter, before which OV-102 would be ready for the first launch. The intention was to build four complete orbiters plus a version for use in structural tests, taking it through the stresses and strains of simulated flights.

The Shuttle was due to be reused many times and it was felt that giving permanent names to the orbiters would be better than calling them Shuttle-1, OV-103 or whatever. President Ford approved of *Enterprise* on 8 September 1976 against the wishes of NASA's hierarchy, who preferred *Constitution* – the roll-out ceremony was even planned for 17 September 1972: Constitution Day. The campaign by 100,000 *Star Trek* fans achieved a somewhat pyrrhic victory, though, because the new *Enterprise* was destined never to fly in space. It was cheaper to convert the structural model (Structural Test Article-099) to space use, with the benefit that it would be lighter anyway because the weight of each successive orbiter was reduced as experience was gained. *Enterprise* would be used only for training and testing.

On 25 January 1979, NASA announced the names of the four flight-worthy orbiters, all from naval ships: *Columbia* (OV-101), *Challenger* (OV-099), *Discovery* (OV-102) and *Atlantis* (OV-103).

Enterprise *was the first orbiter built and is now a museum piece. It is seen here in California being prepared for the captive flight tests. At left is the framework used for mate/demate operations with the jumbo jet carrier aircraft.*

The complete Shuttle stack is transported from the VAB to one of the two launch pads. The white pole on top of the pad's service structure attracts the area's frequent lightning strikes away from the spacecraft. The engine exhaust pit is visible in the concrete base.

By the summer of 1976, NASA was publicly expecting the first test mission with OV-101 to get into orbit at the end of March 1979, with three more tests in July, October and January 1980. As history now records, this schedule – and it was not projecting too far into the future – was wildly optimistic. In reality, the first four tests covered April 1981 to July 1982, some 2 to 2½ years behind what was expected.

The Shuttle Components

Although the Shuttle flies on the wings of modern technology, the basic concept behind the return trips to space is very simple. The three major sections – the orbiter, the external tank and the solid rocket boosters – are put together in the Vehicle Assembly Building of the Kennedy Space Center in Florida, the same building that was used for the old Saturn Moon rockets. The whole Shuttle is then rolled out to the launch pad (the first 24 missions went up from Complex 39A; the ill-fated 25th was the first to use 39B) on the 4500-ton/tonne motorized transporter. The cargo is inserted either while the orbiter is vertical on the pad or, as in the case of the large Spacelab modules, it can be added horizontally before the units are mated together.

A countdown of several days ensues during which the health of the entire vehicle is monitored and the external tank is filled with liquid oxygen and liquid hydrogen propellants at super-cold temperatures. The crew climbs aboard several hours before scheduled lift-off. At launch, the propellants are pumped to the three main engines at the back of the orbiter and the two solid-propellant boosters are ignited to lift the 2000-ton/tonne combination off the pad. The boosters burn for two

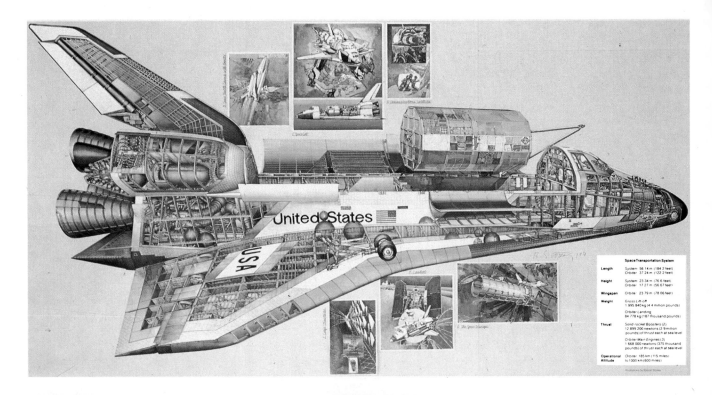

Top:
The Space Shuttle orbiter.

Above:
A fish-eye view of Columbia's flight deck in 1980. The three cathode-ray tubes display flight information to the two pilots; standard flight instruments can be seen on either side.

minutes and are separated when the Shuttle is about 27 miles (44 km) high and travelling at 3100 mph (5000 km/h) They parachute to Earth for recovery and later reuse. The three main engines continue to fire until, at about nine minutes after launch, the external tank is empty and is discarded. Two smaller engines using the orbiter's own small internal propellant supply are then fired several times to establish the type of orbit required for the particular mission. The orbiter is now truly a space vehicle ready to perform its work.

The Orbiter

The most important section of the Shuttle is, of course, the orbiter, designed not only to carry a maximum 65,000 lb (29,500 kg) of cargo into low Earth orbit but also to support up to eight people in space for about a week. Although it is a very different shape, it is about the same size as a DC-9 commercial airliner, some 120 ft (37 m) long, with a wingspan of 80 ft (24 m) and weighing about 150,000 lb (68,000 kg). Most of the structure is aluminium below an outer layer of thermal protection. Moving from the nose aftwards, there is first the flight cabin and crew quarters, followed by the 60×15 ft (18.3×4.6 m) box-like payload bay. At the rear there are the three main engines plus various auxiliary engines for in-orbit manoeuvring.

The flight cabin and crew quarters is the only pressurized section of the whole orbiter, a 2525-ft^3 (71.5-m^3) module suspended and cushioned inside the main structure. Its normal-pressure atmosphere of 79% nitrogen and 21% oxygen is purer than Earth's, with carbon dioxide and other impurities removed by lithium-hydroxide/activated-charcoal canisters kept under the floor of the pressurized section. The oxygen and nitrogen themselves emerge from a regulated system supplied by high-pressure storage tanks under the payload-bay floor just behind the pressurized section. The pressure can be reduced in space in order to shorten the time that intending space walkers have to pre-breathe pure oxygen to clear dissolved nitrogen from their blood.

Flight Cabin

The flight cabin is on top of the crew quarters, appearing superficially like the cockpit of a modern airliner. The two pilots sit in front of five large windows (a great improvement over previous spacecraft) faced by banks of dials, indicators, controls and display screens. They can fly the Shuttle like an aircraft using control columns and foot pedals to control elevons and a large flap at the rear, although these aerodynamic surfaces are useless above the denser layers of the atmosphere. In space, the craft's attitude is maintained by 44 small rocket thrusters located around the nose and the aft section. The orbiter might look like an aircraft but its aerodynamics are far inferior and it is consequently more difficult to fly. The pilots can call on five computers to do the actual flying or they can do it all themselves; there is also a semimanual mode in which the pilot asks for particular manoeuvres and the computers carry them out.

These computers are all-important for a Shuttle mission – it cannot be flown without them. Four are used at any one time, with their decisions being crosschecked to spot any errors, while the fifth comes in as a back-up if one of the primaries is 'out-voted'. The fifth is really for general-purpose work, such as for deploying satellites once orbit has been reached. The first Shuttle launch was delayed in April 1981 because these computers were slightly out of synchronization and a similar problem arose in June 1984 when a microchip failed. These IBM computers are 1972 vintage and new ones should be fitted in 1987 – lighter, more reliable, three times faster and, at $500,000 each, one third the cost of the originals.

At the rear of the cramped flight cabin are controls for doing the important in-orbit work. One of the first jobs in space is to open the graphite/epoxy payload-bay doors to expose the radiators on their inner surfaces. With all of the electronics – computers, radar, communications, etc. – and people aboard a great deal of heat is generated and has to be radiated away by these door panels. If the doors should become stuck then the orbiter has to return to Earth within a few hours to avoid over-heating. During high-heating periods, such as launch and re-entry, a water system is used to absorb much of it by boiling off the liquid into the vacuum of space.

The rear controls are used for satellite deployment and there are controls

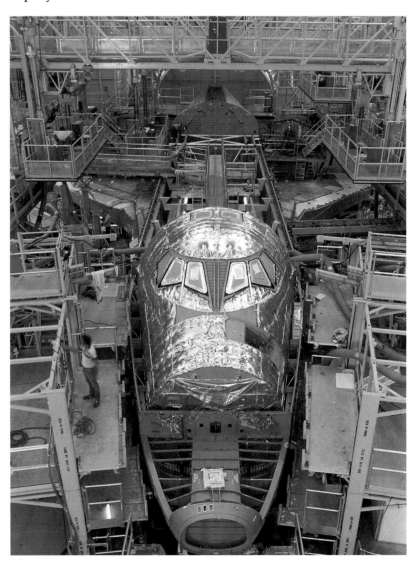

Challenger *under construction. The pressurized crew module is suspended within the outer skin. The wings, minus their leading edges, already have some white tiles.*

Astronaut Ron McNair, killed in the Challenger *explosion, is seen at the rear control panel of the Shuttle trainer.*

similar to those at the front to move the orbiter around during these crucial periods. The astronauts can check on the health of their payloads from this station and they can actually see what is happening in the payload bay through two rear-facing windows. Two overhead windows are useful for observations once the satellites have been sent spinning out into space. One very important set of controls here is used to move the robot arm fixed to one side of the payload bay.

Crew Quarters

The flight cabin is the smaller of the two compartments in the pressurized crew section. Down below is the crew quarters – the living, sleeping, eating and working area that has to provide the comforts of home. On the forward bulkhead are rows of lockers holding carefully-packed food, clothes, personal items, small experiments, tool kits and so on – everything that might be needed for a space mission. The crew enters the orbiter through a 40-inch (100-cm) diameter hatch in the wall directly into this area; the hatch also carries the only window in the crew quarters, a 10-inch (25-cm) diameter porthole.

To the right, against the rear bulkhead, is the toilet area, much beloved of the popular press. Earlier spacecraft (except the roomy Skylab) carried no toilets but forced their occupants to use either tubes for liquid waste or plastic bags stuck on to the buttocks for solids. The new Shuttle version has an inbuilt cup suitable for both male and female astronauts for urinating and a small bowl takes solid waste, sucking it in with a strong airflow and breaking it up with a rotating arm. Unfortunately, it caused problems on all of its early flights (it was not installed for the first test missions) and it had to be sent back to the manufacturer for redesigning. Mission 41D in September 1984 caused embarrassment when urine from the toilet built up: a large icicle around the outlet port on the fuselage had to be dislodged by the robot arm.

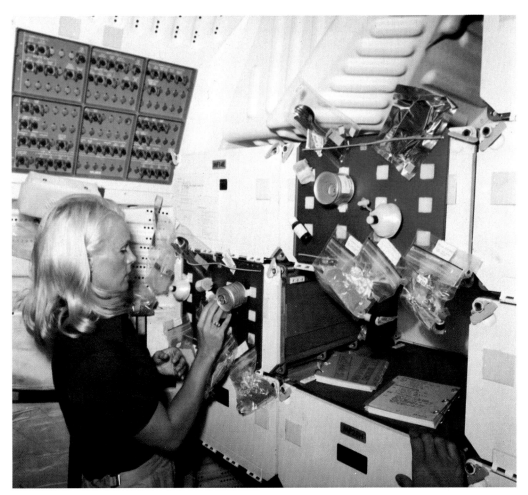

Astronaut Rhea Seddon prepares meal trays aboard the Shuttle trainer in Houston. The lockers contain housekeeping items and small items of equipment but they can be replaced by larger experiment units.

Immediately to the left of the entrance hatch is the galley section, although it is not always carried if there are more pressing experiments. The galley provides hot and cold water, facilities for washing and a small oven to heat food up to 185°F (85°C). Each crew member is provided with 3000 calories per day, spread over three meals repeating after six days. The wrapped meals are kept in the lockers and brought out to be fitted into a tray and heated in the oven or reconstituted with water. Most are either moist or compressed so they can be eaten with spoons or forks without breaking up into crumbs that float off in the weightlessness. Some food and beverages are ready-to-consume from cans.

On the far wall from the hatch is space for several bunks with sleeping bags. Earlier space missions were not so concerned with astronauts' comfort but the lengthy Skylab and Russian Salyut space-station flights demonstrated its importance. As the astronauts are weightless, they can sleep at any angle but the main problem is one of noise. The orbiter is full of plumbing and humming electronics and the occupants sometimes have to wear helmets or ear plugs in order to get to sleep. With up to eight people in a small volume, there is always some activity going on. For emergency

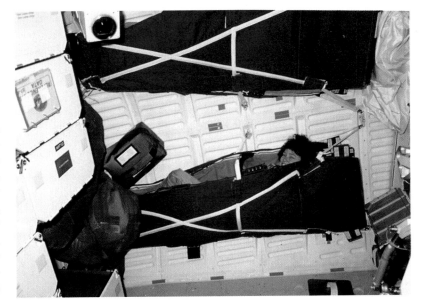

Anna Fisher tests the sleeping arrangements aboard mission 51A in November 1984. Mission 61A a year later tested an inflatable sleeping bag that gripped its occupant to simulate gravity.

flights, the bunks can be removed so that extra seats can be installed to allow an orbiter to carry up to ten people.

Spacesuits

At the rear of the crew quarters is the important airlock used as a doorway into the payload bay for spacesuited astronauts. The spacesuit is an essential part of orbiter equipment. Two or three are now carried on every mission in case mission specialists have to go outside to perform emergency work, such as repairing important thermal protection, in addition to the occasional planned space walks. The Environmental Mobility Unit (EMU), as it is somewhat grandly named, is the latest in a long line of pressure suits stretching back to well before the Second World War. The Gemini suits (1965-66) enabled the first American astronauts to make space walks and those in Apollo (1969-1972) protected astronauts on the surface of the Moon. But, like the rest of the equipment, the suits were built specifically to fit their wearers at great expense. The Shuttle suits are designed to last for at least 15 years and, by swapping the various pieces around, to fit most of the astronauts.

A suit has to carry all of the supplies its wearer needs in leaving the protected environment of the cabin: water, oxygen, heat, radio communications and waste disposal for up to seven hours. An intending space walker first puts on a set of underwear carrying 260 ft (80 m) of tubing to carry cooling water and then attaches a device to collect urine – women have to wear a type of nappy. The leg section, complete with integral boots, is pulled on and then the wearer slides *up* into the rigid fibre-glass upper body section attached by the backpack to the inner wall of the airlock. The pack (or Portable Life Support System as NASA calls it) carries the consumables, plus a 30-minute supply of emergency oxygen, controlled by a microprocessor and controls mounted on a smaller chestpack.

The wearer dons a 'Snoopy' cap carrying microphones and then attaches the gloves, available in 15 sizes. Like the body section, the gloves

Left:
The Shuttle spacesuit is demonstrated by Bruce McCandless during mission 41B. One astronaut always sports red stripes for identification by ground controllers. McCandless is standing on a platform on the end of the robot arm, as used extensively later in satellite repair work.

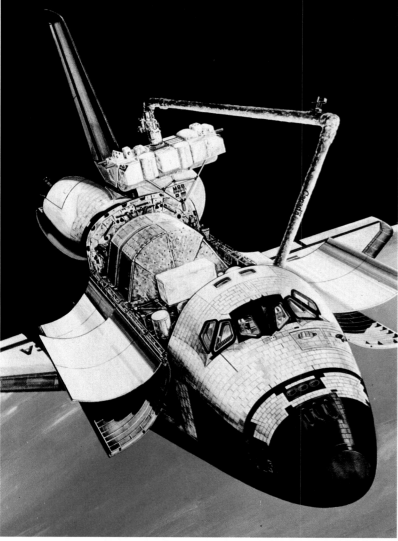

are held on by aluminium ring fittings for speed – the old Apollo suits were distinguished by the time they took to put on because of all the zips, fastenings and the lack of room for manoeuvring. Last of all is the 'fishbowl' polycarbonate helmet and its set of covers and vizors.

Having pre-breathed pure oxygen at the suit's pressure (less than ⅓ of cabin pressure) for up to 3½ hours, the astronaut is now ready to leave through the cylindrical airlock and emerge into the bottom of the payload bay. With a total mass of only 85 lb (38.5 kg), the suit provides protection against heat and micrometeoroids and since two mission specialists always go out together there is cover should an emergency arise. The suit produced some initial problems in orbit that prevented its first exposure to space during STS-5 in November 1982 but since then it has been remarkably trouble-free. Major changes had to be made during its development following a serious fire – always a hazard in a pure-oxygen environment – inside an empty inflated model on 18 April 1980.

Payload Bay

The voluminous payload bay is designed to hold the satellite cargoes and major experiments that cannot fit inside the cabin. The satellites sit on special frameworks reused from mission to mission, covered at launch by the 60-ft (18.3-m) bay doors. The doors are a vital part of the orbiter since they have to be closed completely to keep out the searing heat of re-entry. Normally they are operated electrically from within the cabin but should they become stuck a spacesuited crew member would have to go into the bay and crank a manual system by hand. The first few test missions checked the doors' alignments very carefully since the alternating exposures to the heat and cold of space warps the structures to some extent.

The bay is designed to hold up to four medium-sized satellites and their protective frameworks. A typical launch of a communications satellite would adopt the following schedule.

The clam-shell doors covering the satellite to guard it against the harsh

Discovery's *three main engines, the two medium-powered OMS engines and the smaller thrusters are visible. Each thermal tile carries a serial number. At top right is one set of pad connections into the aft section. Behind the engine bells is the base of one of the solid boosters.*

sunlight and heat are opened, the health of the satellite being monitored all the while by the crew at the rear console in the cabin. An electric turntable then spins the satellite along its main axis to keep it stable at, say, 50 rpm (depending on the requirements) before the satellite is pushed out by springs – the use of rocket thrusters this close in to the orbiter is avoided. The satellite might then drift away for 45 minutes before an onboard timer signals a large rocket motor to ignite and take it to the final orbit. The rocket burn produces a cloud of high-speed gas and particles and the orbiter is carefully turned away so that its high-quality windows are shaded. TV cameras in the payload bay, however, do provide the crew with an adequate view, although once they have released the satellite it is no longer their responsibility.

Running down the length of the payload bay on most missions is the robot arm – or Remote Manipulator System to use its official name. The long jointed arm, controlled from inside the cabin, is used for handling satellites and performing many other jobs that would otherwise have to be done by suited space walkers. The arm can release satellites into orbit, capture others and move equipment and people around the payload bay. In addition, a TV camera can be attached to the 'hand' end so that useful new viewpoints become possible.

Main Engines

Behind the payload bay is the section containing the three main engines and their associated plumbing. The SSME (Space Shuttle Main Engines) are ignited at launch for about nine minutes, drawing their propellants from the huge tank fixed to the orbiter's underside, but are not used again during the entire mission. They are rated at a total thrust of 1.4 million lb (640,000 kg) at altitude but their thrust is varied according to requirements. Consuming 14,100 gallons (64,000 litres) of liquid oxygen and 37,400 gallons (170,000 litres) of liquid hydrogen *per minute*, they might be used at only 65% full thrust to avoid stressing the Shuttle too much during periods of high aerodynamic pressure ('max-q' can sometimes be heard during launch commentaries). At other times they might be throttled up to 104% or 109% of design thrust. It was at this throttle-up stage that *Challenger* was destroyed in January 1986. Each engine is more powerful than the whole Atlas rocket that took John Glenn into orbit in 1962.

These main engines have proved to

Opposite top:
Deployment of a typical medium-sized comsat. The spherical boost motor is attached at the bottom.

Opposite bottom:
The robot arm was used during STS-7 to release and then retrieve the West German pallet satellite.

27

Challenger's main engines are installed in preparation for its maiden flight in 1983.

be a problem area for the Shuttle because not only are they the most advanced cryogenic motors ever built but they are also designed to be *reusable*. Rocket engines take a terrific battering during use, with high-speed turbopumps shovelling supercold propellants into a superhot combustion chamber with pressures up to 200 times greater than the Earth's atmosphere. Usually they are discarded after a few minutes but engineers had to come up with a design that could be used for a total of 7½ hours, or 55 missions, without major overhaul *and* be reliable enough to carry people.

As it has turned out, the engines require considerable work between missions with major components being replaced. Complete engines have sometimes been taken from another orbiter to produce a full working set, and delays have been inevitable. Delays in the test programme led to the anticipated first orbital launch being pushed back, which led to considerable criticism of NASA.

The engines are not the most powerful ever built – the 1½-million-lb (680,000-kg) thrust of *each* first-stage engine in the Saturn Moon rocket was greater than the total thrust of the Shuttle triple cluster – but they are the most efficient for their size. To wring as much power as possible out of a given size of engine it is necessary to pump in propellants rapidly; that is, the combustion-chamber pressure has to be as high as possible. The chamber has to withstand pressures of 3270 lb/in^2 (230 kg/cm^2, or 220 times normal atmospheric pressure) and as the high-temperature exhaust expands through the chamber throat and out into the engine extension, the engine walls are cooled by having cold hydrogen pumped through their tubular construction.

The obvious weak points are the high-speed turbines as the blades are subjected to enormous stresses, but NASA adopted the policy of testing them in engines which themselves were under test. Standard procedure is usually to test components separately and then integrate them step by step. The result was that from 24 March 1977 to 4 November 1979 there were 14 engine test failures, eight of them resulting in fires that damaged the engines.

The test programme has continued in an attempt to increase the engines' reliability, mainly in the area of the high-pressure turbopumps. There were problems with the pumps on most of the initial flights, requiring them to be extensively serviced and repaired. Before the sixth mission in April 1983 *all three* engines had to be removed because of leaks in the hydrogen supply lines. NASA now has a programme underway to improve the engines and intends to spend $1000 million up to about 1995. They might even install new types of engines later to produce more thrust and thus increase the Shuttle's payload capacity.

While the main engines and solid boosters provide most of the power to reach orbit, the final push is given by the two relatively small Orbital Maneuvering System (OMS) rockets housed in two bulbous pods either side of the main engine compartment. These engines are designed to be as reliable and simple as possible and consequently have not produced the

SHUTTLE MAIN ENGINES

Each engine produces enough thrust to power 2½ jumbo jets and if one could be scaled down to weigh only 3 lb (1.4 kg) it would still produce enough thrust to lift a man. The main turbopumps operate at pressures equal to those found 3 miles (5 km) down in the oceans and all of the turbopumps generating at maximum produce more than enough power to propel a battleship.

problems found with their nearby counterparts. Each OMS pod carries its own supply of nitrogen tetroxide (oxidizer) and monomethyl hydrazine (fuel) at normal temperatures to be forced by pressurized helium into the engines, where they ignite on contact. The system, while not suitable for large engines, does do away with much of the complex plumbing and ignition systems. Like the main engines, they

Challenger's *OMS engines are fired during STS-7 to change orbit. At right is the stowed robot arm; the cylindrical containers are Getaway Special canisters.*

Columbia *exhibits missing thermal tiles following STS-3 in March 1982. A technician inspects the vernier thrusters.*

can be swivelled slightly to alter the direction of thrust and each 260-lb (118-kg) engine should last for the entire life of an orbiter (100 missions), allowing up to 1000 ignitions or 15 hours of accumulated firing time. The pods are independent systems but there are crosslines in case one set of tanks or plumbing springs a leak and, for ease of ground processing, the entire pods can be lifted off the back of the orbiter, and pods have been exchanged between orbiters to keep to schedule.

The orbiter does carry yet more rocket engines: 44 of them in all, but these are far smaller than even the OMS set and are thrusters designed for precise manoeuvring in space. Each OMS pod carries 14 thrusters: twelve 870-lb (395-kg) high-power thrusters and two 24-lb (11-kg) low-power vernier thrusters, fed from spherical titanium tanks separate from the OMS supplies although they use the same type of propellants. The upper nose area, in front of the pilots' windows, has 16 thrusters altogether, two of them vernier, in a removable engine section with its own supply tanks. Each vernier motor is designed for up to *500,000* burns, while the most powerful thrusters should survive for at least 50,000. The thrusters are positioned at different angles so that control is possible in all directions. They are even used within the atmosphere during re-entry but only down to an altitude where the aerodynamic surfaces begin to take effect.

A Silica Shield

Competing with the main engines to be the most troublesome part of the orbiter during development was the Thermal Protection System. The aluminium framework is not strong enough to withstand the 3100°F (1700°C) temperatures encountered during re-entry and it has to be protected by a reusable thermal layer that

Thermal tiles are added to Challenger.

does not need much repair work between missions.

The solution was to cover the aluminium skin with silica-based tiles about the size of bathroom tiles (typically 6×6 inches/15×15 cm) that absorb the heat generated and re-radiate it very rapidly. The black areas on an orbiter are the stronger tiles able to withstand up to 2300°F (1260°C) under the body and around the nose area. Around the cockpit windows and stretching back along the sides of the fuselage are areas of lighter white tiles to cope with temperatures of up to 1200°F (650°C). The tiles are such good insulators that you could hold one in your hand even if one side of it were glowing red hot. The low-temperature areas, up to 700°F (370°C), mainly around the top of the payload bay doors and the upper wing surfaces are covered in white, Nomex Felt Reusable Surface Insulation. The hottest areas at the tip of the nose and along the wing leading edges use Reinforced Carbon-Carbon composite made up of graphite cloth soaked in resin.

While most of the black and white tiles have a density much less than water – just 9 lb/ft^3 (145 kg/m^3) – there are about 31,000 tiles on *Columbia*, resulting in a large dead weight of 16,000 lb (7250 kg) that has to be carried to orbit and back. A serious problem arose before the first orbital test flight when it was discovered that some of the tiles might not have been strong enough to survive a mission. They are basically very fragile – they can be crumbled in the fingers – and are held together by outer bonding layers. Thousands of tiles had to be removed from *Columbia* before the first flight and 'densified' by injecting them with a resin; unfortunately, this can take the density up to 22 lb/ft^3 (352 kg/m^3). The later orbiters carry a new type of thermal blanket made from silica fibres on the less sensitive areas to save weight. *Atlantis* has benefited most from this new material.

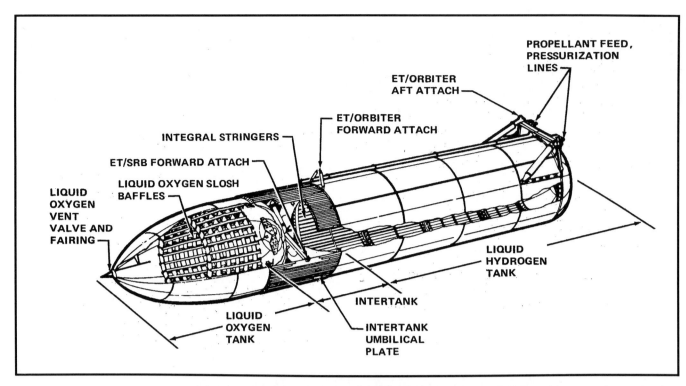

Above:
The external tank.

External Fuel Tank

While the orbiter is a highly complex piece of engineering, the 154-ft (47-m) long and 27.5-ft (8.4-m) diameter $19-million external tank is much simpler. Basically, it holds all of the propellants for the orbiter's three main engines and is thrown away after only nine minutes of flight. Attached underneath the orbiter, it is divided into two sections: the cylindrical hydrogen tank at the bottom holds 226,200 lb (102,600 kg) of liquid hydrogen fuel at −423°F (−253°C) while the smaller pear-shaped oxygen tank at the top can carry 1,359,000 lb (616,400 kg) of liquid oxygen oxidizer at −297°F (−183°C). The exact weights vary from mission to mission depending on the requirements: a light payload requires less propellant.

The hydrogen tank is about three times larger than the oxygen tank because the fuel is so much less dense (16 times lighter). Its storage problems make hydrogen difficult to handle as a fuel but it does provide more energy than using, say, alcohol as did the V-2s.

The external fuel tank is attached to the orbiter at only three locations, with the two rear points providing two 17-inch (43-cm) inlet pipes for the propellants to the engines. When it is discarded 70 miles (110 km) high, a vent at the top is opened to let oxygen gas out to begin a tumbling motion of at least 10 degrees per second. This prevents the tank skimming along the atmosphere and possibly reaching an inhabited area and brings it down instead to break up harmlessly over the Indian Ocean.

The oxygen tank carries anti-slosh baffles around the walls to prevent vibrations forcing the liquefied gas,

Three external tanks, complete with their thermal protection, at Martin-Marietta's Michoud plant in Louisiana.

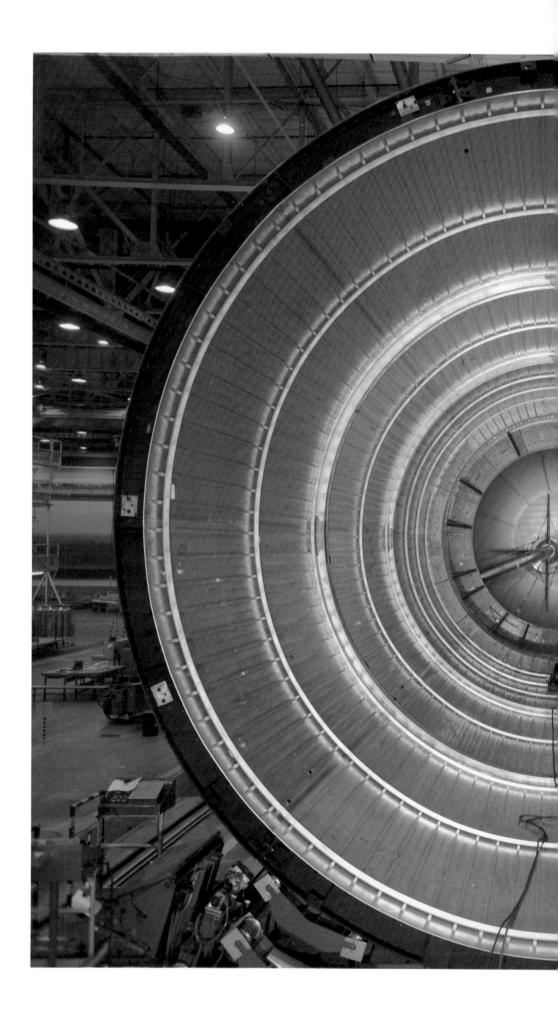

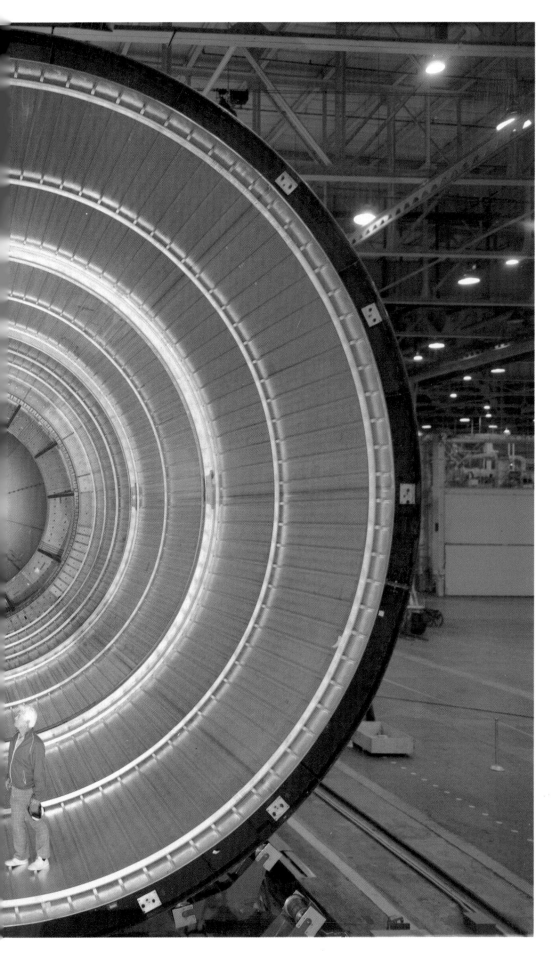

Inside the external tank's hydrogen tank.

which is denser than water, into resonance that could disturb the whole vehicle or even break the external tank away from its supports. Liquid hydrogen is so light that its container does not need baffles. The oxygen leaves its aluminium tank at the base – gravity thus helps to force it out – via a 17-inch (43-cm) steel and aluminium pipe into the intertank area. This area is just a 22.5-ft (6.9-m) high ring used to separate the two tanks and provide attachment points for the solid boosters, with the oxygen pipe emerging through it to run along the outside of the hydrogen tank to an orbiter attachment point. The hydrogen has a shorter route from the base of its tank close to the second orbiter attachment point.

As the external tank carries super-cold propellants and filling operations commence about six hours before launch, it has to be insulated to prevent its contents turning back into gas in the warm Florida climate. Ice could also build up, breaking off at launch and damaging the orbiter's thermal protection. The surface is covered with a cork/epoxy insulation either sprayed on or pre-moulded, depending on the position, which is covered in turn by a polyurethane-like foam to make the thickness up to 1 to 2 inches (2.5 to 5cm).

Since it is carried most of the way towards orbit, the weight of the tank is critical. The first versions weighed about 78,100 lb (35,430 kg) when empty but engineers have brought this down by 10,300 lb (4670 kg) by reducing the thickness of some walls and changing the internal design slightly. Some 600 lb (270 kg) was even saved by dropping the outer coating of white paint!

The Boosters

The solid rocket boosters are reusable and provide much of the power to lift the Shuttle off the pad, with their total thrust of 5.8 million lb (2.6 million kg). Each is designed to survive for up to 20 millions as a new pair costs about $64 million in 1983 terms. Preparing them for reuse (recovery, cleaning and refilling) costs about $25 million, so a saving of some $39 million per launch is made through not using expendable boosters – a saving that the expected Soviet shuttle appears to have sacrificed for higher performance.

The boosters are the largest solid-propellant rockets ever flown and the first entrusted to carry humans; they also hold the record as the largest payloads recovered by parachute. Each is 150 ft (45.5 m) long, the 12-ft (3.7-m) diameter body filled with about 1.1 million lb (502,000 kg) of propellants to be consumed within two minutes. The segments are filled at the Morton-Thiokol company's facility in Utah with a liquid mixture that sets to a consistency of hard rubber. As the steel casings have to survive the 5800°F (3200°C) internal temperature generated during burning, they are first lined with insulation up to 5 inches (12.7 cm) thick, depending on the position, cured in place for 2½ hours in an autoclave. The propellant is a mixture of aluminium powder as fuel (16%), ammonium perchlorate as oxidizer (70%), a small quantity of iron oxide as a catalyst and some 12% of a binding agent to hold the combination together. When the propellant has been cured, the segments can be mated, although this is not usually done until they are required for use at the Kennedy Space Center. Each booster is in four main units, with a parachute container at the top. The base unit carries the nozzle and its steering system, topped by two centre segments and the forward segment with the igniters.

The igniters are small rocket motors in themselves, generating the 5250°F (2900°C) temperature required to set off the main propellants. The propellant inside the booster is carefully shaped with a cavity running down the entire length to control the thrust generated during launch. Liquid engines are easily controlled simply by changing the flow of propellants into the combustion chamber. The drawback with their solid counterparts is that, once started, they burn until they are finished and are not easily controlled. The desired 'thrust profile' is produced by shaping the propellant to a certain cross-section, perhaps with a star-shaped hole or a circle in the centre. A star – the Shuttle's forward segment

The solid rocket booster (SRB).

An extensive test programme certified the SRBs as safe for manned flight.

Each booster is recovered by parachute. This is the left-hand SRB from STS-5 in November 1982.

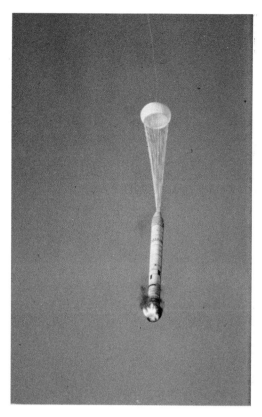

A recovery ship retrieves an STS-2 booster from the Atlantic.

uses an 11-pointed version – exposes a larger burning surface than a circle and will thus produce more thrust over a shorter period. Engineers have designed the boosters so that the ignition flame spreads along the entire length in under a second, producing high thrust rapidly to take the Shuttle away from the pad area. The star gradually burns away so that the thrust decreases at around 60 seconds to prevent the vehicle accelerating too quickly; it then builds up again as the dense atmosphere is left behind. When the insulation layer is reached, the long exhaust plume dies away over about ten seconds and the orbiter's engines are left to complete the job of reaching orbit.

The boosters then fire sets of small solid motors for less than a second at their forward and rear ends so that they simultaneously 'peel' away top-first from the external tank without disturbing the rest of the craft. If the weather is clear enough during a launch, then the separation can often be seen at a height of about 28 miles (45 km), accompanied by puffs of smoke, although their upward momentum can take them up to 44 miles (70 km) before they head back down into the Atlantic. If, as happened with the doomed *Challenger* mission, the boosters fall towards populated areas, they can be destroyed by a radio signal that activates explosive charges. In normal operation, three main parachutes are pulled out of the nose section by drogue 'chutes to lower them into the water but it was found on some early missions that too much damage was being caused by the impact; indeed several sank altogether. The 'chute diameters were increased to 135 ft (41 m) to keep the impact speed below 60 mph (100 km/h) and less damage has resulted as they hit nozzle end first. While the orbiter continues up to orbit, two special recovery ships, the *Freedom Star* and the *Liberty Star*, locate the empty cylinders and pump them full of air for towing back to port. A third recovery ship, *Independence*, will handle the California launches on its own.

The motor segments are sent back to the manufacturers for cleaning and refilling with propellant, while the nose sections are refurbished back at the Cape. Even the parachutes are repaired and reused if possible. The units do not necessarily fly together again as a whole but are swapped around from mission to mission as required.

As with the external tank, engineers have been striving to reduce the boosters' weight. The original empty mass of around 183,000 lb (83,000 kg) was largely accounted for by the 0.5-inch (1.3-cm) steel casing that is designed to withstand the high temperatures and pressures generated within. A 4000-lb (1800-kg) saving has been made by reducing the wall thickness slightly, a feature scrutinized following the *Challenger* disaster. New lightweight casings made from plastic reinforced by graphite fibres began firing tests on 25 October 1984 to prove them suitable for manned flights. Some 33,000 lb (15,000 kg) has been saved on each booster, resulting in 4600 lb (2100 kg) more cargo in the orbiter, although some missions will not need the extra capacity and the heavier steel versions can be used.

The First Flight Tests

It is evident from the preceding pages that the Space Shuttle is a large, highly complex piece of engineering. Each element was tested on the ground – the boosters and main engines were fired many times – but it was impossible to perform a complete test without flying an actual mission. However, the very end of a mission was simulated by dropping *Enterprise* from altitude following its unveiling in September 1976 at Rockwell's plant in California.

The problem of transporting the orbiter high up into the atmosphere before it could be released to make a glide descent was a major one. Previous test craft had used converted

Before Enterprise *was released, several captive flights on the back of its 747 were carried out.*

Enterprise *immediately after its release from the carrier. Fred Haise and Gordon Fullerton piloted it on this first free flight on 12 August 1977. The fairing at the rear of the Shuttle was added to smooth air flow; it was removed for some later test flights.*

B-52 bombers but the new spaceplane was much larger. The solution turned out to be an adaptation of a concept from the early days of aviation when one aircraft was launched from the top of another in flight. NASA bought a Boeing 747 airliner from American Airlines and, having tested the idea in wind tunnels and with flying models, modified it to carry a full orbiter on its fuselage roof. The Shuttle Carrier Aircraft (SCA), as it was now named, would serve not only for these glide tests but also for transporting the various orbiters across the United States on a regular basis during mission preparations.

The 250-ton/tonne combination made its first flight on 18 February 1977 and it was 12 August when *Enterprise* was released to fly on its own for the first time. From a height of 4.5 miles (7.3 km), astronauts Fred Haise and Gordon Fullerton piloted *Enterprise* down like a heavy, unpowered glider towards a runway landing 5½ minutes later at the Edwards Air Force Base in California, just as if they had returned from a mission in space. A varied series of flights that year proved *Enterprise* to be a stable vehicle and the orbital flight tests could be approached with confidence.

But the Shuttle itself is just one component of an extensive space launch system. Almost as important are the launch sites, for there the various elements are brought together for the first time for specific missions. The next chapter discusses the Florida and California starting points for the Shuttle's forays into space.

Leaving the Earth

Cape Canaveral is known throughout the world as the West's major starting point for journeys into space, ranging from near-Earth orbits to the Moon and planets. It is now the home of the Space Shuttle and the many components of the Space Station will converge there for launch in the 1990s. In less than four decades it has been transformed from a swampy scrubland to a high technology spaceport.

The most imposing feature of the whole area is the huge Vehicle Assembly Building (VAB), originally erected to house the Saturn 5 Moon rockets as their three stages and Apollo spacecraft were joined together. For a while, at 525ft (160m) high, 716ft (218m) long and 518ft (158m) wide, it was in volume the largest building in the world; then Boeing's jumbo-jet construction hangar in Seattle, Washington overtook it. It was so big that it was reported to have its own miniature weather system inside, forcing NASA to close the large doors to prevent clouds forming under the roof.

The philosophy behind assembling rockets away from their launch pads is that it provides a more controlled environment, there being no storms or salt spray from the nearby ocean to consider. The other Shuttle launch site, at Vandenberg in California, puts the vehicle together on the pad and then wheels away the protective structure. NASA's method will allow up to 24 Shuttle launches per year, whereas Vandenberg needs to cope with only three or four.

Columbia *is transported along the causeway to Pad 39A from the VAB. The angular building to the VAB's left is the launch control centre.*

Challenger is wheeled into the VAB in preparation for STS-8. The high doors at right were designed to allow complete Saturn 5 Moon rockets to be rolled out.

Inside the VAB. Challenger is lowered to be attached to the waiting external tank and boosters for STS-6.

The main problem with a remote assembly site is that the completed vehicle has to be transported to the launch pad – clearly not a simple job for something weighing thousands of tons/tonnes. The answer is to perform the assembly on a mobile launch platform. The original Saturn-Apollo versions carried large towers to provide support but the shorter, stockier Shuttle can do without and can rest quite happily on its two boosters. First of all, the boosters are built up segment by segment and then the bulky external tank is inserted between them. Only then is the orbiter brought in to be attached at three points to the tank. The platform is then moved by a motorized unit to either of the two Shuttle launch pads.

The major permanent features of the two pads are the towers used to provide access to the vertical Shuttle as it awaits launch. Each pad has a Fixed Service Structure (FSS) and a Rotating Service Structure (RSS) that are essential for preparatory work. An access arm remains attached to the Shuttle until a few minutes before launch in order to provide an escape route for the astronauts. They would have to clamber out of the hatch and slide in baskets down wires to safety.

The actual launch, and the preparations leading up to it, is handled from the Launch Control Center, a low white angular building adjacent to the Vehicle Assembly Building. Like the pads, it has been updated from the Saturn-Apollo era to cope with the modern Shuttle. With the extensive help of computers, the number of people required to sit at desks in two of the four firing rooms (firing rooms 3 and 4 are designed to be 'secure' for classified operations) has been reduced to a fraction of the 450 of the 1960s and 1970s. As the Shuttle sits on the launch platform, thousands of sensors monitor such parameters as engine pressure, tank capacity, liquid levels, turbine speeds, voltages, currents, valve and switch positions and relay all the data through the platform to the Control Data Subsystem on the floor below the firing rooms. From there, it is processed and monitored automatically to pinpoint anomalies in order to warn a human operator. However, the computers can take action immediately to prevent an emergency and have shut down suspect systems prior to launch.

On returning from a mission, an orbiter is towed into the Orbiter Processing Facility, which can accommodate two orbiters simultaneously. It enters a box-like structure (with a cutout above the wide doors to allow the protruding tail to go through!) to be thoroughly inspected. There, the toxic and flammable propellants for the smaller engines are removed and the whole Orbital Maneuvering System pods can be taken off for repair or replacement. If any of the thermal protection covering has been damaged

STS-6 on the pad. The motorized transporter unit has been withdrawn and the crew's access arm with its small room at the end reaches out to Challenger's *hatch. The Rotating Service Structure at left has the TDRS comsat in its large container ready for inserting into the orbiter's payload bay.*

Personnel in the Cape's launch control room no. 1 watch STS-4 lift off. With binoculars is Kennedy Space Center director Richard Smith.

– and some always needs attention – then it can be repaired here. As individual missions carry very different payloads, the payload bay has to be reconfigured for the next mission by removing the old cradles and experiments and inserting the new ones.

Small payloads like the Getaway Special canisters of scientific equipment can be added here, and the large cargoes such as the Spacelab module and its pallets are best installed while the orbiter is horizontal and not on the launch pad. However, satellites to be released in orbit carry their own propellants and small explosive devices, so it is considered safer to install them when the Shuttle is on the pad. Once the orbiter is deemed ready for flight, it is towed the short distance to the VAB to be connected to the other Shuttle elements and then the whole vehicle is transported to the launch pad.

The last major structure relevant to the Shuttle story is not very evident from ground level but it is a welcome sight to returning astronauts: the 15,000-ft (4600-m) long runway, north of the VAB. Positioned on a northwest/southeast bearing, the runway is twice as long and wide as its commercial counterparts and at one end is the Mate/Demate Device to lift an orbiter on or off its jumbo jet for ferry flights.

For orbiter landings, there are sets of microwave scanning beam systems at both ends to tell the pilots and on-board computers the distance, height and angle to touchdown. The double system means that a landing can be made from either direction. Following landing, the orbiter is towed to the OPF along a 2-mile (3.2-km) path.

These, then, are the major features of the Cape Canaveral area. As far as the Shuttle goes, the Kennedy Space Center's responsibility ends when the launch-pad tower is cleared and authority is handed over to the Johnson Space Center in Houston. All of the early space flights were handled from the Cape but from Gemini 4 in June 1965 the actual missions were commanded from Texas. But history might repeat itself and Mission Control could once again be at the Cape as Houston copes with the complex Space Station project.

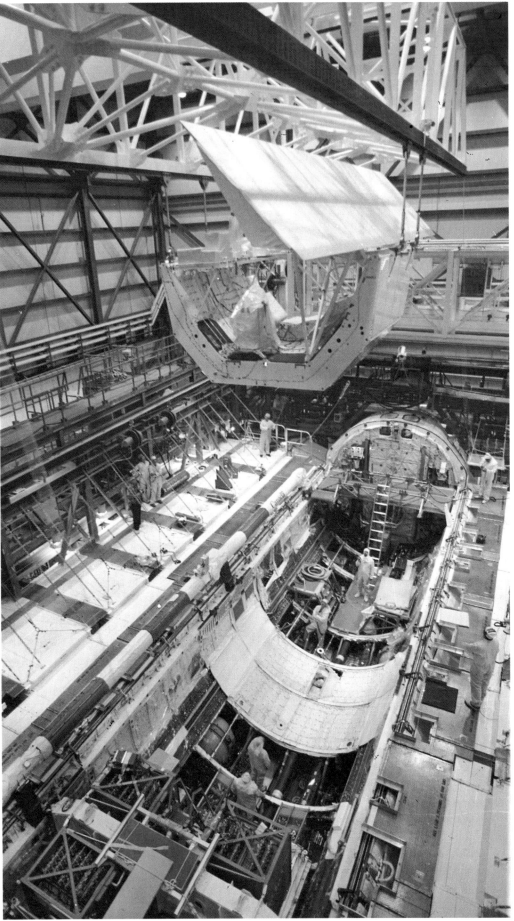

Columbia *is prepared in the Orbiter Processing Facility for STS-2, with the radar mapper being lowered into place. Note the EVA airlock hatch in the crew-compartment bulkhead.*

Vandenberg Launch Site

The Kennedy Space Center will not see all of the Shuttle's operations. Florida is a good launch site for satellites that need to operate in low-inclination orbits (that is, they never stray far north or south of the equator) but a poor base for the polar-orbit operations which many military payloads require. In this way, they eventually pass over every point on the Earth to use their cameras and instruments. Civilian reconnaissance satellites such as weather or Earth-resources craft have similar requirements – the Landsat and Tiros series are good examples. But the problem with Cape Canaveral is that maximum orbital inclination is limited to 57° north or 39° south because of the eastern coast of the U.S.A. and inhabited islands. Flying at angles any steeper to the equator would carry large rocket-propelled potential bombs over populated areas.

The solution to reaching polar orbit is to use the Vandenberg Air Force Base launch site on the coast of California, flying the spacecraft directly south over the Pacific. The base, operated by the U.S. Air Force, saw its first orbital launch as early as February 1959 when the Discoverer spy-satellite precursor became the first man-made object to reach a polar path. Up to the end of 1980, it had seen 478 spaceflights. Vandenberg is similar but somewhat smaller than the Canaveral launch site and the Shuttle will be its first manned project, although a military orbiting laboratory was once planned.

U.S. Air Force officials were aiming for an October 1985 first flight with *Discovery* but the schedule for testing and preparing the facilities became tight. Further delays pushed the mission into summer 1986 and then the loss of *Challenger* put even that target in doubt. When Shuttle flights do begin at Vandenberg, the orbiter will at first still be largely prepared and checked at the Kennedy Space Center before being flown out atop its jumbo jet. But gradually, all responsibility will pass to Vandenberg as the facilities become fully operational and more experience is gained.

Challenger *makes the first Shuttle landing on the Cape's runway, February 1984.*

Launch facilities at Vandenberg Air Force Base on the coast of California. Discovery *was to have been dedicated to this site but* Challenger's *loss may cause revisions.*

Into Space ... Eventually

While the glide tests with *Enterprise* out in California during 1977 demonstrated the orbiter's handling at low speeds, problems with the main engines and thermal tiles threatened NASA's launch schedule. The first orbital flight test (OFT) was then aimed at a March 1979 lift-off but the recurring problems consistently pushed back that first attempt – by more than two years as it turned out.

There was considerable interest within the space community as to who would actually fly the prestigious first two-man mission. Four OFTs would put *Columbia* through her paces in space without major payloads and, if all went well, it would be ready to introduce the era of commercial, reusable space transportation on its fifth launch.

NASA announced on 17 April 1978 that STS-1 (Space Transportation System-1) would be flown by astronauts John Young and Robert Crippen. STS-2 would follow with Joe Engle and Richard Truly; STS-3 with Fred Haise and Jack Lousma; and STS-4 with Vance Brand and Charles Fullerton. All were highly-experienced pilots, although not all had flown in space.

John Young was the natural choice to command STS-1, aiming for June 1979 at the time of the crew announcements. A youthful 50 years of age before *Columbia* actually left Pad 39A at the Cape, Young was the most experienced of the astronaut corps, having made four spaceflights, two of them to the Moon. He had also created something

John Young practises in the flight simulator in Houston.

The Johnson Space Center's flight simulator reproduces the movements and conditions of launch and landing.

Columbia *exhibits her lost tiles following the first ferry excursion.*

of a stir on his first flight in 1965, when it was revealed that he had smuggled a corned-beef sandwich aboard and had offered a bite to his commander, Virgil Grissom!

Young's pilot for the trip was Robert Crippen, at the time without any space experience but by all accounts one of the ablest astronauts. 'Crip' joined NASA in 1969 but when severe budgetary cutbacks lopped missions from the space programme he was left well down the queue of eager fliers. As another Navy pilot, he had been chosen as an astronaut in June 1966 for the U.S. Air Force's Manned Orbital Laboratory which then suffered cancellation in June 1969. Young wanted him along on this test because of his experience in developing the computer operations so necessary for a successful flight.

By the original launch date of March 1979, Young and Crippen were busy practising their mission in computer-controlled simulators while *Columbia* was completing construction in California. It was rolled out on 8 March and towed 38 miles (61 km) by road to Edwards Air Force Base. At this

49

Young and Crippen inspect Columbia's *aft flight-deck station. Note their ejection suits.*

stage it was lacking main engines, OMS pods and about 7800 thermal tiles. The latter, especially, were slowing the work up as each one took about 40 man-hours of work to install – and there were 31,000 in total! The tiles were first glued to a Nomex felt pad to prevent the slight changes in shape of the airframe from cracking them, and then on to a silicone-rubber primer on the aluminium orbiter skin. The new tile then had to be held in place by a support for at least 16 hours to allow the adhesive to cure. The small gaps between every tile were then filled with strips of heat-resistant felt, again to allow for expansion and contraction of the skin.

But this tedious tile work was delaying the rest of the programme and NASA decided to take *Columbia* from Rockwell's plant and ferry it to the Kennedy Space Center where the tiles could be finished while other crucial work – such as installing the engines – could also be carried out.

For the air journey to Florida, technicians added dummy tiles held on by tape in order to present a clean aerodynamic surface but observers were horrified to see hundreds of them fluttering off during a brief test foray into the air on the carrier aircraft. Dismayed technicians counted 4800 lost dummy tiles but, even worse, 100 proper tiles had also gone. If they could not survive a low-speed jet hop, how could they hope to handle a rough journey into space?

The Boeing jet and its cargo finally left to cross the country in several stages on 20 March 1979, reaching the Kennedy Space Center on 24 March. *Columbia* was wheeled into the new Orbiter Processing Facility that day after being lifted off the jumbo's back by the Mate/Demate Device and technicians began the long job of preparing it for STS-1. The external tank arrived on 5 July to undergo processing in High Bay 4 of the Vehicle Assembly Building, awaiting mating with the solid boosters. All the while, an army of helpers was hard at work attaching the rest of the tiles.

The saga of the tiles was not over,

however. Each tile was subjected to a 'pull test' in which a carefully-measured outwards force was applied and the response detected acoustically. Many failed and some even split above the bonding layer – a disastrous situation if high-temperature tiles came off during launch. Altogether, about 4500 tiles had to be removed to be strengthened in some way. NASA was clearly more worried than ever because, on 2 October 1979, they announced that the Martin-Marietta company had been directed to speed up the development of the Manned Maneuvering Unit backpack so that Crippen could wear it to repair any heatshield damage in orbit (the plan was eventually dropped, although the capability is there today).

The time-consuming answer proved to be 'densification' in which the undersides of each tile were soaked with a special compound – thereby increasing its weight – to strengthen them internally. They were then laboriously stuck back on *Columbia* and carefully tested. The process was painstakingly slow and the large work force (about 1400 at peak) took from September 1979 – when the fault was discovered – to 16 November 1980 to complete the job.

The main engine programme was not going well either. When *Columbia's* three engines were installed by summer 1979, the test models at the National Space Technology Laboratory in Mississippi were still having problems. An aborted firing on 4 November showed up a design problem that could have affected *Columbia's* versions and in January 1980 the trio – engines 2005, 2006 and 2007 – were removed for modifications. Later that year, in June, they were back at the test site in Mississippi and each was fired for 520 seconds to prove it ready for flight. All

Columbia's *layer of thermal tiles took years to complete. This 1978 scene shows work on the underside protection.*

Columbia *on the pad with the access arm attached.*

three were reconnected to *Columbia* by 3 August but the story was not yet quite finished – they were taken out *again* on 9-10 October and reinstalled on 8-11 November. There they stayed but there was still a hurdle to overcome: the three had not been fired together as a single unit and NASA management decreed that a 20-second static test should be carried out when the complete Shuttle was sitting on Pad 39A. Not only would it prove the engines but it would also allow the computers and human controllers to practise the launch sequence.

On 3 November the external tank was mated to the boosters in the Vehicle Assembly Building and, at last, on the 24th *Columbia* was rolled out of its nearby processing facility to join them. On the 26th it was upended by cranes in the VAB and attached to the tank and just over a month later the whole combination made the 3½-mile (5.6-km) journey to the pad. The entire vehicle had been checked out during December and a new test series began on the pad to prove that everything was ready for flight. Young and Crippen practised emergency escapes, the tank was test-filled with liquid oxygen and liquid hydrogen and the

Opposite:
A main engine undergoes test firing.

Opposite:
Lift-off! The first Space Shuttle mission begins, 12 April 1981. Never before had a new spacecraft carried humans on its first flight into space.

Columbia *awaits the 20-second test firing of its main engines on 20 February 1981.*

Orbital Maneuvering System tanks were filled with their propellants.

On 17 January the twelfth and final engine cluster test was run successfully in Mississippi and Countdown Demonstration Tests were held with *Columbia* in February, culminating on 20 February with the engines firing for 20 seconds.

Further tests in March with Young and Crippen aboard showed that *Columbia* was ready to begin its 73 hours of final countdown although even at this stage niggling problems pushed the target to Friday 10 April.

The intention was to launch about 45 minutes after dawn to begin a 54½-hour mission during which Young and Crippen would test all of *Columbia*'s basic systems, checking them as thoroughly as possible. Since there was still a question mark over the thermal tiles, the ascent through the atmosphere had been designed to put as little stress on the vehicle as possible.

Young and Crippen slept in the centre's administrative area as the count reached its later stages, being awakened to enter *Columbia* as late as possible. Each event, no matter how small, was listed in the count and had to be achieved on time for the launch to go ahead. Launch Director George Page and his team in the Launch Control Center next to the VAB had to tick off major milestones as they were achieved:

T-14 hr:	Start 2-hr retraction of Rotation Service Structure.
T-5 hr:	Begin final countdown.
T-4 hr 20 min:	Begin filling external tank with liquid oxygen and liquid hydrogen.
T-2 hr 4 min:	Built-in 2-hr hold.

Built-in holds are included so that any problems encountered can be solved without affecting the main schedule. If everything has gone smoothly then the launch crew take a break. In this particular hold, Young and Crippen left their accommodation following the traditional steak-and-eggs breakfast for the van ride to the pad.

T-1 hr 50 min: Young and Crippen climb into *Columbia* and into their seats for launch – Young to the left and Crippen to the right.

At T-20 min there was another built-in hold, this time for 20 minutes. It was at this stage that the launch of STS-1 began to go wrong. During launch and re-entry, four onboard computers are used simultaneously. All must agree in their responses and if one is outvoted by the other three, number five can be brought in as a back-up. The fifth is usually employed in orbit for other work but it was with this unit that STS-1 came unstuck. As the 20-minute hold ended, number five should have come on line to support the others but a timing discrepancy of only 40 *thousandths* of a second was detected between it and the others so the count was halted. The launch could have gone ahead without it but on this first test flight NASA wanted all of the vital systems operational. The count was recycled back to T-23 minutes and held there for 28 minutes before restarting. However, at T-16 minutes, the timing error popped up again and this time the launch was scrapped altogether, disappointing the estimated million visitors crowding the Cape area and countless millions more watching around the world courtesy of TV.

The detailed diagnosis took five hours and since the delay meant that the external tank had to be emptied, a new target 48 hours later was set. It was for Sunday 12 April, *exactly* 20 years to the day since Yuri Gagarin made the first space voyage.

This time there was no problem and the count swept past the T-9 min mark to enter the final sequence. Two minutes later, the astronauts' remaining physical link with the outside world was removed as the access arm began to move away. Until then, Young and Crippen could have escaped down the slide wires on the other side of the service structure, but now they had to rely on their ejection seats. Unlike Mercury and Apollo, there were no special rockets to pull them away from the pad in an emergency and, as in Gemini, they had to rely on being catapulted clear. *Columbia* carried two rocket-propelled seats for the first four test missions only; thereafter they were discarded and replaced with the present operational seats when the ship was re-

turned to Rockwell to be brought up to full operational status.

Once the Shuttle had cleared the tower, Young and Crippen could be hurled clear to descend by parachutes. Overhead panels would first be blown away and the seats would be thrown out from guiding rails along their backs. Since unprotected humans can survive only so far up, the seats could not be used once the 100,000-ft (30.5-km) mark had been reached and the two astronauts would have had to rely on *Columbia* to get them back to the ground.

No Shuttle astronaut now wears a spacesuit for launch or landing but the nature of these test flights and the fact

A Shuttle flight is controlled from this Missions Operations Control Room (MOCR) in the Mission Control Center in Houston. Authority switches from Canaveral to Houston as the Shuttle clears the tower.

that ejection seats were employed meant that Young and Crippen had to wear U.S. Air Force high-altitude escape suits. These five-layer suits weighed only 23½lb (10.7kg) but they provided oxygen and full body protection at altitude. Their orange colour was quite distinctive as the Shuttle spacesuits proper are white – in fact, Young and Crippen each had an EVA suit stowed in the airlock so that they could have emerged to repair the orbiter if necessary.

Young's suit had an interesting feature since, for the first time, an astronaut was about to be launched wearing a pair of glasses. Bowing to the inevitabilities of age, Young had spectacles mounted inside his helmet and wore a pair during orbital operations – several other crew members have since followed suit.

Five minutes before launch, the three Auxiliary Power Units at the rear of *Columbia* roared alive to provide hydraulic power to move the aerodynamic surfaces and 90 seconds later *Columbia* switched to internal power, having previously drawn its electricity through umbilicals attached to the aft area. At T-25 seconds the orbiter's own computers took over the countdown and at T-3.8 seconds the main engines began their start routine, igniting fractionally apart and reaching 90% of full thrust 3.56 seconds later. Hugh Harris,

NASA's launch commentator, counted down 'T minus 10, 9, 8 ... 4 ... we have gone for main engine start' as the barely visible flames built up and steam began to surround the pad from the vaporized water of the pad protection system. In these few seconds, the computers checked engine performance *via* the multitudes of internal pressure and temperature sensors, ready to shut them down immediately before the solid boosters were ignited if they had detected any anomalies. This illustrates why computers are so important: no human could have spotted any problems in the brief period available and the solid boosters would have been ignited, burning until they were emptied.

As it was, the computers allowed the sequence to continue and the boosters ignited at T+2.88 seconds, with explosive bolts firing at their bases to sever the last connections with the pad. A huge cloud of white exhaust immediately sprang up – typical of solid propellants – and the 222,894-lb (101,104-kg) *Columbia* leapt away from Pad 39A just 3.98 seconds past 7. a.m. local time to begin the first Shuttle mission. There was no turning back and the computers kept a watch on engine performance ready to signal a flight abort if necessary. If two engines had failed before seven minutes had elapsed then Young and Crippen would have been thrown clear but the more likely loss of a single engine would have forced them to execute an RTLS (Return to Launch Site) abort after shedding the two boosters. As *Columbia* flew over the Atlantic, it travelled with the astro-

Recovery of an STS-1 booster.

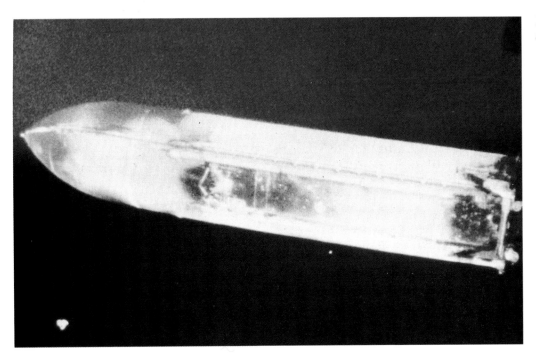

Ejection of the STS-1 external tank.

nauts' heads pointing almost *down* so that in an RTLS abort it would have gradually flipped over completely and returned to Florida. This was a complex manoeuvre and one to be avoided if at all possible. If there were sufficient speed, *Columbia* could press on across the Atlantic and land at the prepared Rota U.S. Naval Air Station in Spain. If an engine failed just before orbital speed was reached, then *Columbia* could have flown around the world and landed in California – about 52 hours before it was supposed to.

Climbing on almost 6½ million lb (3 million kg) of thrust, *Columbia* soared over the Atlantic into the blue sky. The trajectory was about 5° steeper than planned and the booster engines were providing more thrust than expected; at one stage astronaut Dan Brandenstein, acting as 'capcom' in Mission Control-Houston, told them '*Columbia*, you are looking a little hot, all your calls will be a little early', meaning that they would be reaching events in the flight path earlier than planned. As it happened, the two factors almost cancelled each other out and there was no problem.

At the 132-seconds mark, the solid boosters had done their job and were ejected to splash into the Atlantic five minutes later for recovery, leaving the Shuttle with 5.8 million lb (2.6 million kg) less thrust. It was 31 miles (50 km) high and moving at about 2900 mph (4670 km/h) some 30 miles (48 km) down-range from its starting point as the main engines continued to gulp propellant from the external tank.

Columbia passed Mach 4 after about 4 minutes and at 6 minutes 40 seconds, while travelling at 15 times the speed of sound, it rotated to head slightly *downwards* – but still flying east – from 84 to 72½ miles (135 to 117 km) high to pick up speed. Brandenstein told them 'You are single engine press to MECO', meaning that they could still reach orbit if there was only one engine firing at Main Engine Cut-Off. The dive took *Columbia* to Mach 25 and the climb began again as the engines were pulled back to only 65% of full thrust – there was plenty of capacity now. Brandenstein warned them to expect engine cut-off at 8 minutes 34 seconds, and it came right on time as *Columbia* travelled at 17,502 mph (28,161 km/h) in a temporary 15×93 miles (24×150 km) orbit. The tank was discarded, an onboard camera later showing how charred it was in some areas, to descend over the Indian Ocean and be destroyed. Crippen had exclaimed at one point 'Man, that was one fantastic ride'. His heartbeat had reached 130 per minute while veteran Young had stayed at a cool 85. He said that he was excited too 'but I just can't make it go any faster'.

Top:
The dark patches are thermal tiles missing from Columbia's pads. This is a TV view transmitted by Young and Crippen to mission control.

Above:
A view from STS-1: the Cape Cod area of the U.S. eastern seaboard on 13 April 1981.

Over the next seven hours, Young fired the two Orbital Maneuvering System engines four times to reach the final orbit of 170×172 miles (274× 277km), but after the second burn 44 minutes into the flight the major job had been to get the payload-bay doors opened. This was because the inner surfaces were used to radiate away waste heat and if there had been any problems the OMS burns would have been used to bring *Columbia* down after only five hours instead of reaching the higher orbits.

Up to this point, the crew could not see the aft portion of the orbiter because the two rear-facing windows were blanketed by the gloom of the payload bay. But as Crippen carefully opened the doors it became evident that something was amiss – he could see dark patches on the front parts of the OMS pods where thermal tiles should have been. TV pictures sent down from *Columbia* showed the gaps quite clearly and the media immediately began to sensationalize the story: would *Columbia* survive the searing heat of re-entry with 15 or 16 tiles missing? The answer – as the astronauts and engineers well knew – was yes because the tiles were not in critical areas. What was more worrying was the possibility of tile loss on the underside where the belly and wings were protected by the high-temperature black tiles and where Young and Crippen could not see. On later missions the robot arm and its TV cameras would be able to inspect this area by craning over the side but STS-1 did not have this luxury; the crew had had their hands full with training for everything else and an arm was not carried.

The critical under tiles had all been 'densified' and well tested after bonding so NASA's engineers were not unduly worried and it is possible that spy-satellite cameras were turned towards *Columbia* to photograph it. Some reports say that this was done by at least one KH-11 satellite while others have suggested they were too far away – but NASA and the U.S. Air Force are not saying!

Nevertheless, Young and Crippen were pleased that the doors had worked well, Crippen noting that they seemed 'to hesitate' slightly before opening – a characteristic not encountered in simulations on Earth.

As everything was going well, the pair took off their orange suits some 3½ hours after launch as they flew over the U.S.A. for the third time and settled down to a well-deserved meal. The next few hours were busy as OMS burns three and four put them into their proper orbit, followed by a long series of tests with all of the smaller thrusters spread around *Columbia*'s nose and tail to check that they performed as expected. Young occupied the left-hand seat, as normal, checking switch settings against a detailed list and wearing spectacles almost in the half-moon style. He noted that the larger thrusters sounding like 'muffled howitzers' firing. He was pleased that the control system had '...gone as smoothly as it could possibly go. All the ...jets have been fired; the vehicle is performing like a champ.'

Another meal followed and, after a few further chores, Young and Crippen settled down after 13 hours of continuous activity since lift-off to get what sleep they could. On subsequent flights, the crew would be able to sleep in more comfortable conditions but, since this was a new spacecraft, they stayed in their seats to be ready for any emergency. It was no great hardship as every American astronaut apart from the Skylab-dwellers of 1973/4 had had to sleep in similar conditions. This was the first time, though, that one could wake up and be greeted by such a wonderful view through larger, near-panoramic windows.

The pair complained of being cold during the night – the problem was a heating valve – but during day two they completed a full programme of work. More thruster tests and TV transmissions came and went, including a talk with Vice-President George Bush, and Crippen closed and reopened the doors again to check if they had warped during the 1½-hour day and night cycles. They passed with flying colours and there has been no significant problems with the doors in any mission since then.

During launch, Crippen referred to eating a corned-beef sandwich – a dig at Young's practical joke during Gemini 3 in March 1965 that had earned him a rap over the knuckles from the NASA hierarchy. During their 21st circuit of the world, the astronauts put their ejection suits on again and clambered into the seats as a rehearsal for re-entry the next day. Before going to sleep for the second and last night, they performed an alignment of the Inertial Measurement Unit in the nose, as they had to do several times a day. The IMU was there to tell them exactly where and in what direction they were pointing by taking account of all the manoeuvres made beforehand. It was vitally important for it to be accurate – especially for re-entry – and it was constantly checked and updated by navigational sightings on stars. Young made it one of his last jobs before the sleep period and the first in the morning.

The next day, during the 34th circuit after about 49½ hours of orbital flight,

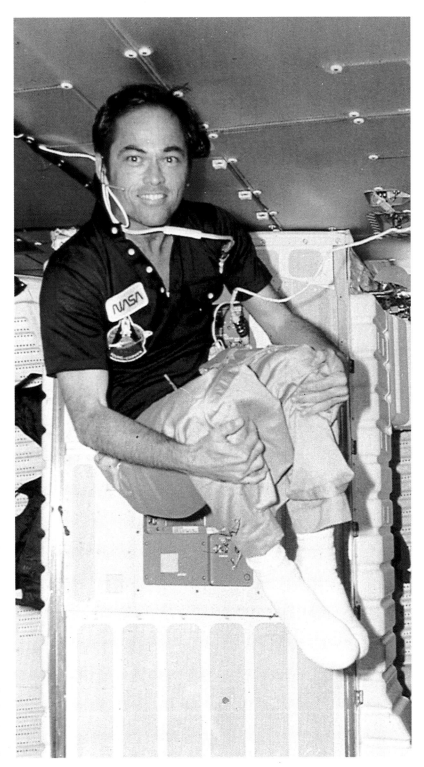

Crippen in Columbia's *mid-deck area.*

61

Columbia *rests on the lake bed at Edwards Air Force Base with service vehicles approaching.*

the pair put on their suits for the last time and closed the doors using the controls at the rear of the cabin. They climbed into the forward seats and Young rotated Columbia so that the OMS engines at the back faced the direction of motion. On pass 36, as they flew over the Indian Ocean 53 hrs 21½ mins into the mission, the two 6000-lb (2720-kg) thrust OMS engines roared into life for 160 seconds to bring Columbia's path within the atmosphere by chopping 200 mph (320 km/h) off the orbital speed.

The new, brief orbit brought the spaceplane into contact with the atmosphere over the eastern Pacific as it headed north to begin the long glide to California for touchdown at Edwards Air Force Base. As Columbia ploughed into the denser atmosphere with its nose angled up 40°, the thermal tiles began to glow and Young and Crippen were presented with a pink glow out of their windows. Contact was lost for 16 minutes when the temperature rose high enough to ionize the surrounding air, thus blanking out radio signals, but the effect was well-known from previous space missions. They emerged from blackout 35.6 miles (57.3 km) high, radioing down 'Hello Houston, Columbia here, we're doing Mach 10.3 at 188 (188,000 ft)'.

Young ordered the computers to fly a series of S-shaped curves to bleed off speed as California approached, crossing the coast at Mach 6.6. Columbia passed over the runway 10 miles (16 km) up and the hundreds of thousands of spectators were shaken by a double sonic boom as the orbiter dropped below the speed of sound.

Columbia was now moving on the Heading Alignment Cylinder, spiralling down an imaginary cylinder in the sky to line up for the 20° glide down to the runway. All of the complex manoeuvres had been designed to bring Columbia down to the surface of the 7-mile (11-km) stretch of dried lake bed at about 215 mph (346 km/h) – as a glider, it would not get a second chance to land.

The landing gear came down a few seconds before contact and the rear wheels touched first, followed by the nose dropping. Joe Allen, the astronaut in contact with Young and Crippen, exclaimed 'Welcome home Columbia. Beautiful!' The first Space Shuttle mission was over.

By 29 April, Columbia was back in the processing facility at the Cape to prepare for STS-2 that September. Some 300 tiles were damaged – most only slightly – but it became clear that the new launch date would not be met: the two weeks turnaround time between flights was still far in the future.

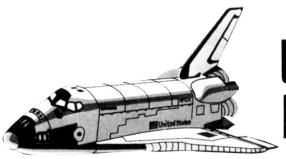

Up and Running

The glorious success of STS-1 began the Shuttle orbital flight programme on a high note. Could subsequent flights equally live up to expectations?

The major feature of the Shuttle is that it is designed to be a regular ferry into space, with as little refurbishment as possible between missions. STS-1 had subjected *Columbia* to a benign flight regime but the second launch would provide a more rigorous test, as well as carrying experiments in the payload bay and testing the robot arm for the first time.

Piloting the first spacecraft to go into orbit twice would be Joe Engle and Dick Truly, both new to space, although Engle had flown the X-15 rocket aircraft in the 1960s. NASA was aiming for a 30 September 1981 launch but an accident on the 22nd ruled that out when corrosive nitrogen tetroxide being loaded into the forward thruster tanks leaked and damaged several hundred thermal tiles. It was just this sort of incident that had to be avoided if NASA was to demonstrate rapid Shuttle turnaround.

The 73-hour countdown began on 31 October but, just 31 seconds short of ignition, the computers called a halt as they detected low pressures in the electricity-generating fuel cells carried at the bottom of the payload bay. Following other problems with Auxiliary Power Units, the 4,473,133-lb (2,029,000-kg) STS-2 eventually left Pad 39A at 3.10 p.m. GMT on 12 November 1981 – Dick Truly's 44th birthday.

Columbia had been in orbit for only 2½ hours when one of the three 200-lb (91-kg) fuel cells began giving trouble again. Mission controllers in Houston decided to take no chances even though a single cell was enough to run all of *Columbia*'s electrical equipment. Astronaut Sally Ride radioed up to Engle and Truly barely a day into the flight: '*Columbia*, this is Houston. We have some bad news for you. You're going to be coming home tomorrow.' They were clearly disappointed the five-day mission was being chopped down to 54 hours: 'Oh boy, that's not so good'.

Before landing in California, however, they were able to test the robot arm, finding that it behaved much as

Joe Engle enjoys an exercise period on the mid-deck treadmill.

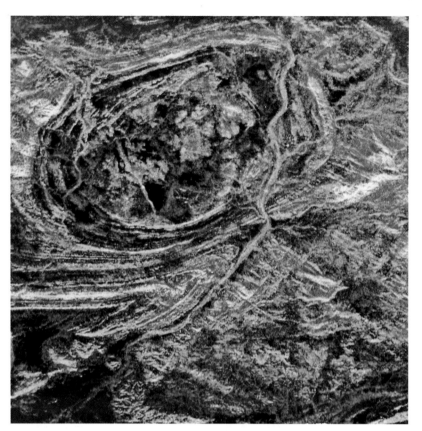

expected. The other experiments, such as the radar for mapping the Earth, were also cut short but still returned valuable data. The radar surprised scientists by revealing unsuspected ancient river beds below the arid Sahara.

The Tests Are Completed

Despite the early end of STS-2 – forced more by ground rules than by technical trouble – *Columbia* had performed well. Flights STS-3 and STS-4 still remained before full operations could begin and were designed to push the spaceplane's abilities even further.

STS-3 became the first to launch on the targetted date – although an hour late – taking Jack Lousma and Gordon Fullerton into orbit on 22 March 1982 for a seven-day mission. Their task was to give the robot arm a more thorough workout and test *Columbia*'s response

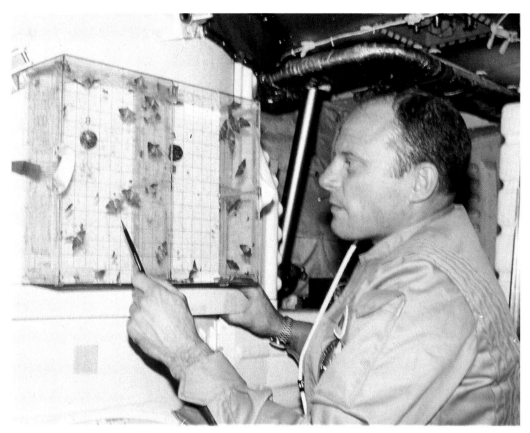

Opposite top:
The radar aboard STS-2 highlighted topographical variations in western Australia's Hamersley mountains.

Opposite bottom:
Columbia *returns to Earth.*

Left:
Lousma inspects Todd Nelson's moth/bee flying experiment.

Below:
Jack Lousma (left) and Gordon Fullerton piloted STS-3. Behind Lousma is the cylindrical container of the Monodisperse Latex Reactor.

SPACE MANUFACTURING

STS-3 carried test versions of important space-based manufacturing equipment. The Monodisperse Latex Reactor (MLR) grew minute polystyrene spheres from smaller seed spheres in a latex mix. On Earth, gravity interferes and the spheres emerge angular and in a variety of sizes. The tiny spheres are used in medical research – for measuring the size of pores in the eye for glaucoma research, for example, or as carriers of minute quantities of drugs. Batches of space spheres went on the market in 1984 while others were held back for subsequent flights to grow larger ones.

The electrophoresis unit promises greater medical advances. It can separate out cells, hormones and enzymes from biological mixes very accurately as they flow through an electric field, yielding, on STS-4, purities four times better than possible on Earth and at 700 times the quantity. This commercial venture led to McDonnell-Douglas company engineer Charles Walker flying three times to operate the equipment in the Shuttle's cabin. A larger unit was due to travel aloft in the payload bay in 1986 to prepare for commercial production of new, purer pharmaceuticals.

65

The Plasma Diagnostics Package is moved around Columbia *by the remote manipulator.*

to harsh thermal conditions by flying with one side facing the Sun for long periods. Much time was devoted to experiments, one of them designed by High-School student Todd Nelson to see how moths and bees would cope with flying in weightlessness. Housed in the cabin for the first time were the Monodisperse Latex Reactor and electrophoresis experiments to demonstrate the benefits of zero-gravity manufacturing, while the payload bay carried the 10,000-lb (4540-kg) OSS (NASA's Office of Space Science) unit housed on a pallet built by British Aerospace. Instruments were included to observe the Sun and to collect tiny meteorites whizzing through space.

The failure of the arm's TV camera at the grappling end cancelled the attempt to lift the 800-lb (360-kg) Induced Environment Contamination Monitor out of the bay and the astronauts had to be content with moving the 353-lb (160-kg) Plasma Diagnostics Package around. IECM was intended to measure the pollution around *Columbia* – leaking gases, water, dust and so on – because cleanliness was a concern for future payloads. Astronomical telescopes have to avoid making their observations through clouds of debris and the trend now is towards such instruments flying away from the orbiter. The PDP was successfully shunted around recording electrically-charged particles in the vicinity and testing the handling of the arm in the process.

It became evident there was a problem with the thermal tiles again when Lousma noted some missing from the nose section and remnants were picked up around the launch pad. Possibly ice breaking off the external tank had damaged the orbiter – indeed, the astronauts had commented on the amount of debris flying around during launch. None of the lost pieces exposed critical areas but post-flight analysis revealed that 1033 had to be either densified or replaced altogether.

Apart from Lousma's space sickness, the loss of three out of four communications channels and the toilet breaking down again, STS-3 uncovered no major difficulties. The Californian landing site had been waterlogged before launch and NASA decided to use the Northrup Strip in New Mexico instead of the concrete runways at Edwards or the Kennedy Space Center. High winds precluded a landing after seven days – the crew were told just 30 minutes before the de-orbit burn with the OMS engines – and touchdown had to wait until four minutes after the eight-day mark.

When STS-4 took off on time on 27 June 1982 to become the 83rd manned mission in orbit, with Tom Mattingly and Henry Hartsfield aboard, it had been agreed between NASA and the Department of Defense that no TV coverage of the secret U.S. Air Force equipment in the payload bay would be broadcast. Neither would newsmen hear any of the astronauts' conversation with ground control during sensitive operations.

In fact, *Columbia* was housing the

Commander Mattingly in the mid-deck area during STS-4. Below his head is the MLR unit.

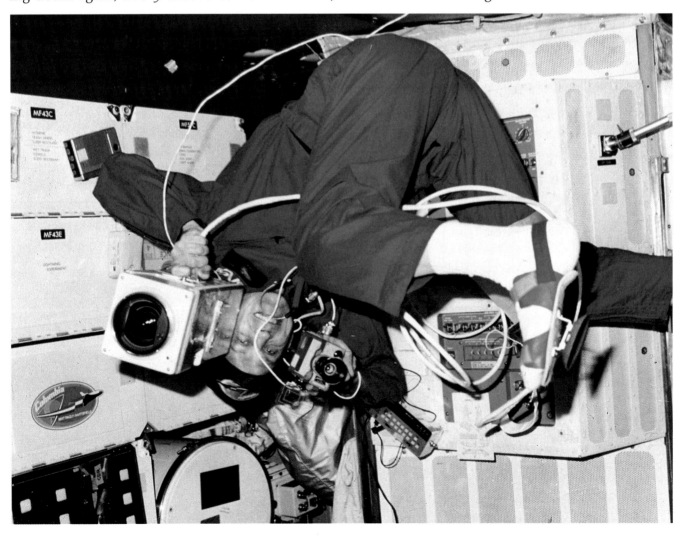

Cirrus experiment, designed to study the atmosphere in infrared with a telescope cooled down to −436°F (−260°C) by liquid helium. The work would be useful in building new detectors for early-warning satellites to detect heat (infrared) exhaust from missiles and aircraft.

Cirrus became something of an embarrassment when the telescope cover refused to move and even attempts to nudge it off with the end of the robot arm came to nothing. This was the second major blow to the mission as both solid boosters sank to the bottom of the Atlantic following parachute failure after separation from *Columbia*. However, the rest of the mission was more fortunate. The electrophoresis unit and Monodisperse Latex Reactor were aboard again and this time, unlike STS-3, the heavy Induced Environment Contamination Monitor was lifted around the payload bay on the end of the arm.

The bay also held the Shuttle programme's first fare-paying passenger: a Getaway Special canister. This small container, offered by NASA to educational, research and commercial enterprises for up to $10,000, held nine experiments from the University of Utah looking at the effects of weightlessness on materials processing and living matter, pioneering a use of cargo space now made by dozens of other investigators.

Mattingly tried on a spacesuit in the airlock for the first time on day five in preparation for the first space walk during the next mission. Earlier, *Columbia* had been turned to heat its underside in sunlight because engineers thought that water might have seeped under the tiles during a thunderstorm the day before launch. If it turned to ice in orbit, it could have cracked off some of the protection.

Mattingly guided *Columbia* to a precise landing after 7 days 1 hour on the concrete runway at Edwards Air Force Base on 4 July in front of a large Independence Day holiday crowd. President Ronald Reagan personally welcomed the two men but disappointed space watchers by not committing the U.S.A. to the expected Space Station project in his speech.

That announcement had to wait until January 1984.

'We Deliver'

With testing of the spacecraft completed, the Shuttle programme now opened for business. The fifth Shuttle flight had one major commercial objective: the delivery of two fare-paying communications satellites to orbit. The

The returning astronauts are greeted by President and Mrs. Reagan.

crew were also to test the Shuttle suits in open space for the first time. When pilots Vance Brand and Bob Overmyer took off on time on 11 November 1982 they were accompanied by mission specialists Joe Allen and William Lenoir in the first ever four-man launch. Henceforth, pilots would concentrate on flying the orbiter while in-orbit work such as using the robot arm and deploying satellites would be handled by mission specialists. Later flights also included payload specialists who were not professional astronauts but selected by scientists to operate specific experiments aboard the spacecraft.

Both Lenoir and Allen had been waiting for space missions since 1967 but budget cutbacks had seriously affected their scientist-astronaut group and only now with the Shuttle was it

Joe Allen undergoes the indignity of medical tests.

being fully employed. The pilots occupied their normal positions but their ejection seats had been disarmed and they no longer wore high-altitude suits. Lenoir sat on a folding chair at the back of the flight cabin, while Allen had to be satisfied with the rather restricted view below in the mid-deck area below the flight cabin. All four wore blue NASA coveralls and helmets that could provide emergency oxygen for short periods.

Behind Allen in the payload bay were the SBS-3 and Anik C3 communications satellites housed in protective cradles. The mission specialists began checking them several hours prior to release using the displays and controls at the rear of the cabin, before Overmyer turned *Columbia* so that the payload bay faced in the appropriate direction. The SBS sunshield was opened and for the first time the crew and TV viewers could see the cylindrical satellite, with its thousands of solar cells, and folded dish antenna on top. Both satellites were of the HS-376 type built by the Hughes company at a cost of $25 million each and launched by NASA for $9 million. NASA was offering attractive rates to tempt users from expendable launchers; but on top of that there was the cost of each satellite's solid-propellant Payload Assist Module boost motor.

A turntable rotated SBS up to 50 rpm to provide stability after release and Lenoir sent the command to sever the connection with *Columbia*. Small explosive bolts fired to cut the holding clamp and springs pushed the satellite out at 2 mph (3 km/h). It rose slowly, glinting in the sunlight with the dark blue of the solar cells contrasting with the gold of its thermal insulation. TV cameras watched it float away and Overmyer moved *Columbia* some 16 miles (26 km) and turned it to protect the windows as an onboard timer signalled the PAM motor to ignite and take SBS towards geostationary orbit 22,300 miles (35,900 km) high. Once there, a smaller motor fired to raise the low point of the orbit, causing the satellite to circle the Earth once every 24 hours so that it remained in position over the United States to provide commercial radio, TV and data links. At separation it had weighed 7225 lb (3277 kg) but in final position its weight had dropped to 1394 lb (632 kg). The two satellites and their cradles had weighed about 14,600 lb (6620 kg) at launch – well below *Columbia*'s payload capacity but this was only the first cargo mission.

The next day, during orbit 22, Allen released Anik and that, too, performed perfectly, taking up position as part of Canada's domestic communications system.

The crew were clearly delighted as they showed TV viewers a card reading 'Ace Moving Company. Fast and Courteous Service. We Deliver.' The smiles faded somewhat when the next major job, the space walk, was tackled. The schedule called for Allen and Lenoir to move into the airlock and don

Above:
SBS-3 springs out of its protective cradle. Behind is the closed sunshield of the Anik C3 comsat.

The crew celebrates their successes. Holding the card is Vance Brand, at left is Bill Lenoir, top is Bob Overmyer, right is Joe Allen.

Challenger *lifts off on its maiden voyage, 4 April 1983.*

Opposite:
Technicians gently mate NASA's large TDRS comsat with its IUS upper stage before insertion into Challenger. *At top are the folded antennae.*

the suits during orbit 46 on day four to begin 3½ hours of pre-breathing pure oxygen to wash the nitrogen out of their blood before they breathed the suits' reduced-pressure oxygen. They would then emerge into the payload bay.

But it was not to be. First the walks were delayed after Lenoir suffered a bout of space sickness and then, as Allen twisted and turned his way into a suit inside the cramped airlock, a fan motor for circulating oxygen began to fluctuate. Lenoir's suit joined in with low oxygen pressure. Later analysis on the ground pinpointed a faulty sensor in Allen's case, while wrong assembly during manufacture had ended Lenoir's hopes.

The failures were not serious as space walks were not yet required for actual in-orbit work. The main thing was that the two satellite deployments had been successful and when *Columbia* touched down on a concrete run-way at Edwards AFB after 5 days 2¼ hours, NASA was already looking forward to the next mission.

STS-6 was another landmark for the Shuttle because a new orbiter, *Challenger*, would make its debut. It was 2488 lb (1128 kg) lighter than *Columbia*, mainly because engineers had been able to shave mass off the structure as they gained experience during the *Enterprise* and *Columbia* test programmes. Some 600 lower-temperature tiles were replaced by a lightweight thermal-blanket material and all 30,000 of the others had been densified to avoid the troubles that had plagued *Columbia*.

Challenger was delivered to Florida on 5 July 1982 but two test firings of the three main engines in December revealed leaks and the entire set was replaced. The launch was further delayed when high winds drove sand from the nearby beach into the payload bay and technicians took the payload – NASA's own $100-million Tracking and Data Relay Satellite (TDRS) – out for checking and cleaning.

Mission six began at 6.30 p.m. GMT on 4 April 1983 carrying pilots Paul Weitz and Karol Bobko and mission specialists Story Musgrave and Don Peterson. For the first time, one of the new external tanks weighing some 10,000 lb (4540 kg) less than earlier models was used and the main engines were burned at 104% of original design thrust most of the way to orbit.

Once in space, Story Musgrave took control of the long series of tests to check out TDRS before its planned early departure. Just ten hours after launch, he released the 5000-lb (2270-kg) satellite from its cradle raised 59° above the horizontal and watched it glide silently above the windows in the roof of *Challenger*'s cabin. Fifty-five minutes later, the first stage of the U.S. Air Force Inertial Upper Stage rocket carrying TDRS-1 fired and successfully took its payload into a temporary orbit, leaving the astronauts to go to sleep satisfied with a job well done.

Unfortunately, as they slept, the second stage ignited and after 80 seconds of the planned 105-second burn contact was lost with the satellite as it tumbled out of control. Months of

Musgrave (with red bands) and Peterson undertake the first Shuttle space walk. Behind Musgrave is the vacated IUS holding cradle.

subsequent analysis showed that heat from the motor nozzle had caused part of its support to collapse, leaving it pointing to one side. A long and complex story in its own right, ground controllers subsequently managed to use the satellite's own small thrusters to nudge it gradually into its proper position in geostationary orbit by 29 June.

Challenger and her crew were completely blameless and it was not until 1985 that the second IUS was used aboard a Shuttle after thorough investigations and modifications. NASA's TDRS-2 had to wait until the ill-fated 51L mission.

On day four, Musgrave and Peterson cavorted for 3½ hours in the now-empty bay, testing the suits, tethers, tools and general Extra-Vehicular Activity (EVA) procedures. Everything went beautifully and the pair returned from the first American EVA for nine

---

SHUTTLE SPACE WALKS
1. STS-6, April 1983. Musgrave and Peterson. Basic tests.
2. 41B, February 1984. McCandless and Stewart. Test of MMUs. First free-flying human (McCandless).
3. 41C, April 1984. Nelson and van Hoften. Repair of Solar Max. Use of MMUs.
4. 41G, October 1984. Leestma and Sullivan. Refuelling satellite tests. First U.S. woman space walk.
5. 51A, November 1984. Allen and Gardner. Recovery of Westar and Palapa satellites. MMUs used.
6. 51D, April 1985. Hoffman and Griggs attach 'fly swats' to RMS in unscheduled walk for Syncom rescue.
7. 51I, August 1985. Van Hoften and Fisher repair Syncom satellite during two walks.
8. 61B, December 1985. Spring and Ross test space construction methods during two walks.

years to the congratulations of their crewmates.

Day five, like all penultimate mission days, was spent preparing for re-entry, clearing equipment away, installing the extra seats and setting the computers. *Challenger* returned to Edwards just 24 minutes into the sixth full day after 80 orbits, with Weitz allowing the computers to do the flying until 4900 ft (1500 m) before touchdown. The new spaceplane proved to be in excellent shape, requiring relatively little work before STS-7 two months later.

Challenger suffered no major delays and began the next mission on time on 18 June with an interesting mix of experiments and commercial payloads. Bob Crippen, becoming the first man to fly the Shuttle for a second time, commanded the first crew of five, among them the first American woman astronaut, Dr. Sally Ride.

With them were pilot Fred Hauck and mission specialists Dr. Norman Thagard and John Fabian, the doctor added at a late stage to concentrate on space-sickness studies. Up to half of all space travellers fall sick for the first few days in orbit and on Shuttle sorties of a week it can affect the work schedules. So far there has been no way of predicting who is susceptible and Thagard's job was to observe himself and his colleagues – although this crew stubbornly remained healthy!

Ride and Fabian released the $24-million Canadian Anik C1 and the $37-million Indonesian Palapa B1 comsats during the first two days. The commercial record was now four out of four.

Days three and four were mostly spent working on experiments. The electrophoresis and Monodisperse Latex Reactor units were carried yet again in the mid-deck area and the payload bay held no less than seven Getaway Special canisters with experiments from schools, industry and laboratories. The bay also contained the West German SPAS-01 – standing for Shuttle Pallet Satellite – for release and then retrieval by the arm to demonstrate satellite reusability. It also had a camera that showed viewers on Earth how the Shuttle appeared in orbit.

Above:
The first five-person space crew. From right: John Fabian, Sally Ride, Fred Hauck, Bob Crippen and Norman Thagard.

Left:
Sally Ride operated the electrophoresis unit during STS-7. The liquid flowed inside the clear-fronted box at left.

Below:
The SPAS-01 satellite took this photograph of Challenger *in orbit. The arm has been crooked in the shape of a 7 to symbolize the mission number.*

Crippen had been due to bring *Challenger* down to the first landing on the new runway at the Cape but, despite a one-orbit delay, fog persisted in the area and the 97-orbit, 6 day 2½ hour flight ended at Edwards after shifting 850 miles (1367km) crossrange in the atmosphere to reach the alternative site. A wingless spacecraft could not have done that.

The First Night Launch

STS-7 was a resounding success but STS-8 found NASA in something of a dilemma. It should have carried TDRS-2, the second of the agency's large comsats, but rectifying the Inertial Upper Stage's faults was proving to be time-consuming and the Shuttle was left with a near-empty payload bay. True, there was the small Indian Insat-

Above:
The Payload Flight Test Article was used during STS-8 to test the robot arm's reaction to large masses.

Right:
STS-8: the first night launch of the Shuttle programme.

Dr. Bill Thornton spent most of his time in orbit studying space sickness. At extreme right is the airlock hatch.

1B weather/communications satellite and four Getaway Special canisters but on their own they hardly justified a Shuttle flight. NASA therefore decided to bring forward the 7462-lb (3384-kg) Payload Flight Test Article (PFTA) from STS-11 for testing the robot arm's response to very large masses. It was really just a deadweight with five grappling points at various locations but its length of 19.8 ft (6.03 m) gratifyingly filled much of the payload bay.

When STS-8 lifted off on 30 August 1983 it was the first manned night launch since Apollo 17 headed for the Moon in December 1972 and once again the surrounding area was bathed in the glow of a man-made Sun. The five-man crew of commander Dick Truly (ex-pilot of STS-2), pilot Dan Brandenstein, and mission specialists Dale Gardner, Guion Bluford and Bill Thornton had a busy time ahead of them in orbit. Bluford, the first black American in space, began work straight away with the electrophoresis unit as it carried living cells for the first time in four outings. He spent most of the day testing the separation and purification of chemicals from live human pancreas, kidney and pituitary gland cells, while Dr. Thornton concentrated on space-sickness studies. Like Thagard on STS-7, he had been added to the existing four-man crew to study the problem and he reported 'I think I learned more in the first hour and a half of flight than I have in all the previous years. I'm convinced the problem is solvable.' Such studies continue to this day.

On day two, Insat-1B was deployed during *Challenger*'s 18th circuit, heading towards its geostationary position over the equator at 74° east longitude. Its PAM boost motor performed as expected but the American-built satellite itself caused trouble when its solar panel did not want to deploy properly.

The next three days were spent in tests with the PFTA, with Gardner lifting it out of its seating while the pilots fired *Challenger*'s small manoeuvring thrusters to see how the 50-ft (15.2-m) long arm reacted. It remained stable.

The night launch dictated a night landing at Edwards and Truly brought *Challenger* down on the brightly-lit dry lake bed on 5 September to complete the six-day trip. Despite the near-perfect mission, NASA was criticized for flying it without a major cargo, the detractors pointing at the 260,000 postal covers, included because of the spare capacity, as trivial users of a very expensive launch system. Nevertheless, it did allow a launch schedule to be maintained at a time when the space agency was eager to demonstrate reliability.

Europe In Space

Challenger had flown three very successful missions and it now took a rest while *Columbia* reappeared to fly the major STS-9/Spacelab 1 mission.

One of the largest cargoes carried in the Shuttle is the European-built Spacelab. NASA and the European Space Agency agreed in 1973 to build a pressurized laboratory that would fit into the Shuttle for use by scientists in orbit. ESA would foot the bill (ending up at about $1200 million) and NASA would provide the ride into orbit for the European experiments and astronauts.

Basically, Spacelab consists of sets of modules 8.9 ft (2.7 m) long and 13 ft (4 m) in diameter connected to the Shuttle's airlock by a tunnel. Scientists can then transfer into the normal atmosphere of the modules and work at experiments mounted on racks. The modules can be joined together to produce a double unit with 4600 lb (2100 kg) of research equipment. Experiments that need direct exposure to space are mounted on pallets built by British Aerospace, 9½ ft (2.9 m) long and 13 ft (4 m) wide, and designed for 50 flights. Some Spacelab missions will not require a pressurized section at all and will use up to five pallets along the length of the payload bay.

Spacelab 1 was designed to test the laboratory concept thoroughly with 71 experiments covering five scientific

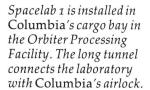

Spacelab 1 is installed in Columbia's *cargo bay in the Orbiter Processing Facility. The long tunnel connects the laboratory with* Columbia's *airlock.*

The Spacelab pressurized module and its connecting tunnel as seen from Columbia's *rear windows.*

disciplines: astronomy and solar physics, life sciences, atmospheric physics and Earth observations, space plasma physics, and materials science and technology. The intention thereafter was to devote subsequent Spacelabs to more specific areas of research; Spacelab 2 in 1985, for example, concentrated on astronomical observations.

Coping with the packed schedule was the first six-man crew commanded by the most senior astronaut of all, John Young. His pilot was Brewster Shaw, flying for the first time, while the others were all scientists: Owen Garriott and Robert Parker as mission specialists, and non-astronauts Byron Lichtenberg and West German physicist Ulf Merbold.

Columbia had undergone extensive changes to prepare it for the 33,252 lb (15,083 kg) laboratory. The three main engines had been removed for refurbishment and installation on the fourth orbiter, *Atlantis*, and replaced with engines 2011, 2018 and 2019 capable of burning at 104% of rated thrust. A new antenna had been added to relay the huge amount of data through TDRS-1 and a galley unit added forward of the hatch.

The completed vehicle was rolled out to Pad 39A on 28 September 1983 on schedule for a 28 October launch but an unpleasant discovery made on the recovered left-hand booster from STS-8 ended any hopes of a late autumn departure. Engineers found that insulation on the nozzle interior had nearly burned away, almost allowing the 5600°F (3100°C) exhaust to break through. While NASA denied that it could have been fatal, checks revealed that STS-9's right booster was in a similar condition. The stack was returned to the VAB for the offending booster aft unit to be replaced. This incident was quickly recalled at the time of the *Challenger* disaster.

Having returned to position on 8 November, *Columbia* and its precious cargo lifted off on 28 November and began flying along the eastern seaboard of the U.S.A. instead of directly out over the Atlantic. This was because some of the experiments, such as Earth observations, required as much of the surface to be covered as possible and a 57° inclination, the maximum allowable from the Cape, would reveal most of Europe. For the first time, the Shuttle was visible over Britain, a rapidly moving dot brightly illuminated by the Sun from below the horizon.

Just four hours later, the hatch leading into the 19-ft (5.8-m) linking tunnel was opened and Garriott, Lichtenberg and Merbold happily floated into the laboratory under the watchful gaze of TV cameras.

For the next ten days the astronauts operated the 38 instruments (20 in the module, 16 on the pallet and two divided between both) in an orgy of science. For example, telescopes

Garriott draws blood from Lichtenberg for later analysis on Earth.

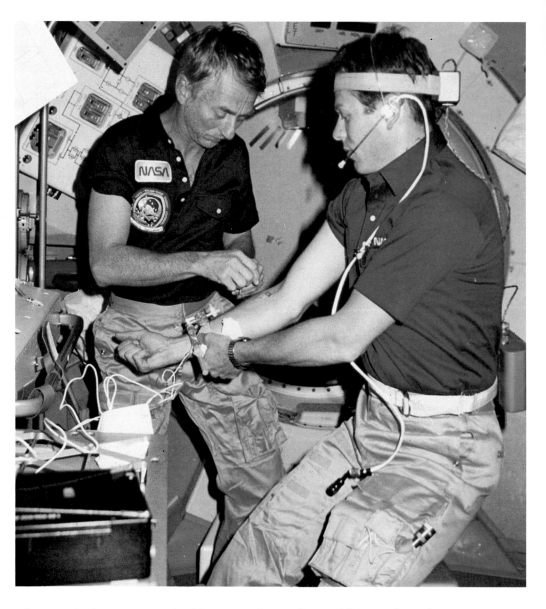

Opposite:
Spacelab carried a camera for taking high-quality Earth photographs through a window in the module's roof. The film magazine jammed but Bob Parker made a temporary darkroom out of a sleeping bag and saved the project. This picture shows western China on 3 December 1983.

observed the cosmos in X-ray and ultraviolet, although some of the UV photographs were inexplicably fogged, and observations of the Sun were made to see if the amount of radiation emitted by our nearest star is varying appreciably. If it is – and recent work tends to suggest that it is – there could be important implications for the Earth's future climate.

On 8 December, Young and Shaw were preparing for *Columbia*'s descent when they were startled by one of the forward thrusters jolting the craft more than normal and computer 1 failed. Minutes later, computer 2 similarly went down. Young later said 'When the first one went, my knees shook. When the second went, I turned to jelly.'

The second unit was brought back on line within an hour but computer 1 stubbornly stayed out of commission and the re-entry was delayed while ground engineers studied the problem. Post-landing inspection revealed there to be microscopic debris trapped inside the integrated circuits. Young was eventually given the go-ahead to re-enter without computer 1 and he brought *Columbia* down to a perfect landing on circuit 166 after a mission of 10 days 7 hours 47 minutes.

But the drama was not yet over. Leaking hydrazine from two of the three Auxiliary Power Units in the rear ignited and started a small fire minutes before touchdown. If all of the APUs had been knocked out early on, Young would have had no control. As it was, the fire burned itself out in minutes and there was no danger.

Reaching Maturity

Bruce McCandless tests the manoeuvring precision of the arm in preparation for the Solar Max repair on the next flight.

1984 was intended to introduce the reusable spaceplane into regular use. No less than 11 flights were included in the ambitious programme, mostly carrying comsats, with the exciting capture and repair of an ailing astronomical observatory in prospect. As it turned out, the actual flight record bore little resemblance to what was planned – only five missions left the ground. Two TDRS-carrying launches were dropped because the IUS boost stage was still

not ready, a problem that spilled over into the military flights, and others were shifted around to accommodate the changes.

Nevertheless, 1984 saw a dramatic series of space sorties, with the last one of the year rescuing two satellites lost from the first. Mission 41B, with commander Vance Brand, pilot Bob Gibson and mission specialists Ron McNair (later to die in the same Shuttle), Bob Stewart and Bruce McCandless, got off to a perfect start on 3 February 1984, but it soon turned sour after the commercial Westar 6 comsat was released from *Challenger* on orbit six. The astronauts completed their part without flaw but when the satellite's PAM boost motor fired it burned for less than ten seconds instead of the planned 85, leaving it in a 219×873 mile (352×1404 km) orbit instead of the intended 190×22,236 mile (306×35,778 km). The same happened with the second comsat aboard, Palapa B2, on 6 February.

It turned out after lengthy investigations that a batch of PAM rocket nozzles were faulty and the pair just used had burned through.

Much of the rest of the flight was in preparation for mission 41C in April when two astronauts would repair the Solar Max observatory in orbit.

McCandless tests the manoeuvring backpack.

The MMU provides a new degree of freedom for astronauts in orbit.

McCandless and Stewart ventured into the payload bay to fly the Manned Maneuvering Unit (MMU) for the first time – a 300-lb (136-kg) backpack powered by 24 nitrogen gas thrusters for sorties away from the orbiter. McCandless became the first free-flying human satellite and he and Stewart demonstrated a special device attached on the front for docking with Solar Max.

The EVA successes made up for the

earlier disappointments and *Challenger* capped a busy mission by making the first landing at the Cape on the concrete runway after almost eight days in space.

Repairing Solar Max

When the $77 million Solar Maximum Mission satellite was launched by a Delta rocket on St. Valentine's Day 1980, it carried seven instruments to observe the Sun in unprecedented detail at a time of maximum solar activity. As the Sun's radiation provides the energy to drive our planet's weather, Solar Max's results are directly applicable to climate research. Unfortunately, after only nine months, the satellite's attitude control system blew three small fuses and the loss of the fine-pointing capability meant that four telescopes were essentially out of commission.

One of NASA's stated objectives for the Shuttle is for it to act as a celestial breakdown service and this opportunity was just too good to pass up. Not only that, but to build and launch a replacement observatory would have cost $235 million, as opposed to $40 million for the repair work. NASA was not reticent about pointing out the saving.

Challenger headed for the highest orbit of the series so far on 6 April 1984, commanded again by Bob Crippen, making his third trip. With him were pilot Francis Scobee (later to die in command of the same orbiter) and mission specialists Terry Hart, James van Hoften and George Nelson. After reaching an initial orbit of 132×290 miles (213×467km) to begin the chase for Solar Max, they released the 10-ton/tonne Long Duration Exposure Facility that occupied most of the bay. This was a 12-sided framework housing a variety of simple experiments for retrieval from orbit by mission 51D in February 1985. They are still up there!

Day three began the major part of the mission. Crippen, Scobee and the computers brought *Challenger* to within 200ft (60m) of the slowly-turning Solar Max, glinting in the sunlight. Nelson and van Hoften immediately moved into the payload bay – time was short since the manoeuvres meant that *Challenger* had little propellant to spare – and Nelson donned an MMU together with the docking device attached to the front of his suit. Pulsing his small gas thrusters to leave the orbiter at 0.5 mph (0.8 km/h), he spent the next eight minutes drifting over to the satellite. The plan was for him to dock with a small trunnion pin on its body, cancel out the 1° per second rotation and hold it steady for Hart to grab with the robot arm. Solar Max was one of the first satellites to be built with retrieval in mind so no trouble was expected.

However, it was a complete failure. Nelson matched the satellite's rotation and carefully moved in between the solar panels to dock – but the jaws did not snap shut. He tried three times but each time he just bounced away. Nelson later said that a nearby small thermal insulation fastener was protruding, preventing his docking device from pushing all the way home.

The Long Duration Exposure Facility was released with experiments requiring lengthy exposures to space. Visible behind the arm's wrist joints is Cape Canaveral.

George Nelson attempts to dock with the SMM satellite on 8 April 1984. The solar telescopes observe through the apertures on the end face.

Grabbing a solar panel with his hand made the tumbling worse and a backup plan using the robot arm also failed. Ground controllers managed to get the satellite stabilized again over the next day and on day five Terry Hart latched on to it with the arm at the very first attempt. He lowered it into a special cradle in the bay and Nelson and van Hoften were able to make all the repairs and replacements in a 7 hour 18 minute space walk.

Solar Max was released the next day, 12 April, and Crippen brought *Challenger* back to California after almost seven days to a very warm welcome.

More Delays

Major assembly work had started on the third orbiter during the second half of 1980 at Rockwell and delivery had been made to the Cape on 9 November 1983. *Discovery*'s external finish was noticeably different because the new lightweight blanket material replaced most of the white tiles on the upper surfaces, resulting in a basic weight of 147,925 lb (67,100 kg).

The three main engines were test fired on the pad for 18 seconds on 2 June 1984 and on the 25th commander Henry Hartsfield, pilot Michael Coats, mission specialists Richard Mullane, Steven Hawley (Sally Ride's husband) and Judith Resnik (a subsequent *Challenger* victim), together with payload specialist Charles Walker, were aboard when computer 5 (the back-up) failed. With a replacement from *Challenger*, the next day's attempt ended dramatically just four seconds before booster ignition when the computers spotted a fuel-valve malfunction as the first main engine began burning. Working at superhuman speed, the computers signalled shutdown as the third engine was about to start and began to make

the vehicle with its explosive propellants safe. A fire started in the engine area from leaking hydrogen fumes and mission controllers were considering an emergency egress as water jets sprayed the area. Hartsfield elected to stay put and the crew were brought out 40 minutes later as yet another new orbiter demonstrated teething troubles. The centre engine with the faulty valve was replaced on 5 July.

41D went aloft almost on time on 28 August and the first major job after opening the doors was for Resnik, the fourth woman in space, to test out the new robot arm. The SBS-4 commercial comsat was released later in the day and the first use of a PAM upper stage since 41B in February 1984 went perfectly. The similar Telstar 3 followed it two days later. The previous day, a new type of comsat had been deployed to act as a link for U.S. Navy forces around the world, leased at $17 million per year from its builders, the Hughes company. Its 14-ft (4.3-m) diameter body took full advantage of *Discovery*'s bay width, rolling out sideways at 2 rpm in

James van Hoften stands on the arm platform to open up the side of the SMM.

Syncom 4-2 rolls out from the rear of the payload bay. In the foreground is the solar-experiment storage box. Note the camera on the stowed arm's elbow joint watching the satellite deployment.

'frisbee' fashion. Syncom 4-2 also carried its own large kick motor internally so no PAM-type stage had to be added. The three deployments earned NASA just $37 million at a time when it was being estimated that each Shuttle mission cost four or five times that amount.

On the second day, Charles Walker began the first of his planned 80 hours on the electrophoresis equipment. Walker was a McDonnell-Douglas engineer intimately involved with the processing project and NASA had not only allowed the unit to be carried for nothing but it also allowed an expert operator aboard, having charged only $80,000 to cover his basic training costs. The company was working towards new commercial pharmaceutical products and Walker was there to process as much material – the nature of which was then still secret – as possible. The drug was later identified as erythropoietin, a hormone used for combating anaemia and with a potential market of several hundred million dollars a year. It is not widely used at

Charles Walker attends to his company's electrophoresis unit. The red bands in the unit show the separation process underway.

the moment because current processing methods do not remove by-products harmful to humans. The purer space version should be available before the end of the decade.

With the three satellites gone and the first Shuttle passenger hard at work, the remaining 41D major objective was the testing of a large simulated solar array. On days three and four Resnik extended 84 panels from a 7-inch (18-cm) high package to various heights up to their full 102-ft (31-m) length. A real array of this type, with a width of 13 ft (4m), could produce 12.5 kW of electricity from thousands of solar cells and it was necessary to study deployment techniques and how such a large flimsy device would behave once extended. *Discovery*'s thrusters were fired and the sail's motions carefully recorded for later analysis but the astronauts there and then declared themselves delighted with its response.

As 41D wound down, the crew were embarrassed by new additions to *Discovery*: two large icicles from the toilet and waste-water outlets built up on the port side and the worry was that they could break off during re-entry and damage crucial tiles. Commander Hartsfield took the responsibility of using the arm himself in a tricky manoeuvre close to sensitive areas and he succeeded in snapping off most of the offending pieces, thereby ending the media's interest in the flight.

89

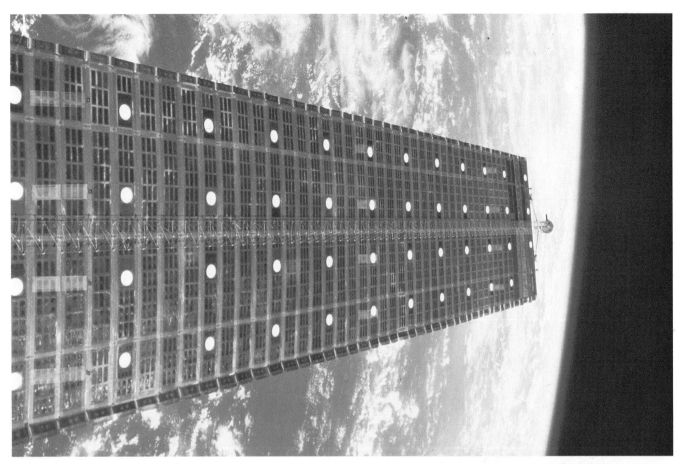

As *Discovery* landed in California on 5 September, NASA was aiming to get two missions off the pad over the next two months. On recent performances, it was a tall order.

Challenger's was the first with a record crew of seven for mission 41G. Commander Bob Crippen (his fourth space trip), pilot Jon McBride, and mission specialists David Leestma, Sally Ride and Kathryn Sullivan were joined by two latecomers: Canadian Marc Garneau and U.S. Navy oceanographer Paul Scully-Power. Six Canadians had trained for two or three places made available by NASA in return for that country's contribution of the robot arm to the programme; Scully-Power would be the first trained oceanographer to view the globe from such a vantage point. In fact, most of 41G's work was devoted to Earth observations and Scully-Power observed several phenomena, notably currents, that no one had recognized before.

Launch on 5 October 1984 was followed by Sally Ride releasing *Challenger*'s only satellite: the $40-million, 5090-lb (2310-kg) Earth Radiation Budget Satellite, which would observe the amount of solar radiation received and then re-emitted by the Earth, for climate studies.

While Scully-Power worked alone watching the Earth below the 57° inclination orbit, the rest of *Challenger*'s crew was concerned with the Shuttle Imaging Radar. A more powerful version of SIR-A carried by STS-2 in November 1981, SIR-B sent radar pulses down to the surface and used the return signals to build up images showing details as small as 82 ft (25 m) across and penetrating cloudy regions. Over the next few days there were problems as the Ku-band dish antenna to the right of the cabin would not point properly at the TDRS-1 satellite to relay the vast quantity of data being collected. The crew had to lock it in position and bodily move *Challenger* to direct it at NASA's comsat. It was a cumbersome process and it was not helped by the radar's large flat antenna refusing to latch down whenever major manoeuvres were made – Sally Ride

Opposite top:
The large simulated solar array was carried to study the erection and behaviour of such a flimsy panel.

Opposite bottom:
Greece and numerous Aegean islands are visible in this on-board picture from Challenger.

Marc Garneau (left) and oceanographer Paul Scully-Power in front of the airlock.

David Leestma practises in-orbit satellite refuelling techniques.

Opposite:
Dale Gardner rescues Westar 6.

Discovery To The Rescue!

While there had been little during mission 41G to excite public attention, the rescue of the Palapa and Westar comsats during 51A created worldwide interest. When the two satellites were lost after successful deployment from 41B the previous February, their London insurance underwriters had to pay out $180 million to Westar and the Indonesian government. The underwriters now effectively owned the craft and by paying just $5.5 million to NASA for the actual rescue and $5 million to Hughes for preparatory work, they could have the satellites refurbished and sold to recoup some of the lost money.

It was a daring venture and 51A commander Frederick Hauck put the chances of full success at only 50%. Launch on 8 November took *Discovery* into an orbit slightly lower than its targets so that it could gradually catch up while other cargo was released.

Palapa and Westar had been brought down into paths about 224 miles (360 km) high by firing their own kick motors and small liquid thrusters. The errant PAM motors had already been released.

On days two and three the Anik D2 and Syncom 4-1 comsats were released by mission specialists Joe Allen, Dale Gardner and Anna Fisher; pilot David Walker was the fifth person aboard.

By day five, *Discovery* had been brought to a halt a short distance from Palapa as Allen and Gardner hauled themselves out of the airlock. Allen fitted himself into a backpack and attached what NASA called a stinger device to his front. The problem with Palapa and Westar was that neither, unlike Solar Max, had been built with rescue in mind so there were no grappling points for the robot arm. Allen approached Palapa and pushed the stinger inside the nozzle of the burntout boost motor and released small arms that sprang out to complete a hard dock. Using his backpack's thrusters, he brought the 1226-lb (574-kg) satellite over to *Discovery*.

After a struggle with badly-fitting

had to hold it down with the arm during OMS burns to prevent it flexing around.

On day seven, Leestma and Sullivan emerged for a 3½-hour EVA, Sullivan becoming only the second woman to venture into open space – Svetlana Savitskaya had become the first only four weeks earlier from the Soviet Salyut 7 space station.

Leestma concentrated on demonstrating refuelling techniques for the Landsat 4 Earth-resources satellite, making connections and penetrating a practice plumbing system. The pair manually stowed the stubborn radar and Ku-band antennae safely for re-entry and rejoined their colleagues to spend day eight preparing for descent.

handling frames, Palapa was locked down in the payload bay in a cradle. Following a rest day, it was Gardner's turn to capture Westar, leaving it nestling snugly next to its partner ready for the journey home on 16 November.

Even the space programme's critics had to admit it was a magnificent achievement. Unfortunately, a growing surplus of comsat channels caused the new owners difficulty in finding buyers for their 'secondhand, as new' satellites.

The Military Aspect

The much-tested and modified Inertial Upper Stage finally made its Shuttle reappearance with mission 51C in January 1985, beginning the year's schedule with a secret military payload. The all-military crew of commander Tom Mattingly, pilot Loren Shriver, mission specialists Ellison Onizuka (later to fly on 51L) and James Buchli, together with U.S. Air Force payload specialist Gary Payton, had originally been assigned to the defunct STS-10 a year earlier.

Even some of the most basic details of the three-day mission were kept secret by the Pentagon, such as the exact launch time, and NASA officials were ordered to stay quiet. Most of all, the satellite being carried by *Discovery* remained hidden but experienced space-watchers could deduce some facts. The use of the powerful IUS meant that it was a large payload going into a high orbit, so it was possibly either an early-warning craft or some kind of comsat. *The Washington Post* was not revealing a major secret when it reported in its 19 December 1984 issue that the satellite was for 'signals-intelligence' purposes, listening to radio and radar emissions over the Soviet Union.

Discovery was launched on 24 January and Gary Payton was responsible for checking the satellite, later reported to be of the Acquacade type, before it was released some 16 hours after lift-off. The IUS first stage fired 55 minutes later, producing slightly lower thrust than expected, and the second stage completed the journey towards geostationary orbit. No TV pictures or radio transmissions of the release were made public and the successful IUS transfer was not announced until 16 hours before Mattingly guided *Discovery* down in California after only 3 days 26 minutes in space.

The press, naturally, reacted against the Pentagon's restrictions, comparing them with NASA's usual open-handedness. The civilian space agency was somewhat embarrassed but up to a third of all future Shuttle flights will be for military purposes and NASA, like everyone else, had to toe the line. These restrictions have applied for years to unmanned military launches and no one had complained; the only difference now was that a manned spacecraft was involved.

The Mission That Nearly Was

Following 51C, the next up should have been 51E the next month. It was to have carried the heaviest cargo load yet (53,400 lb/24,200 kg), in addition to a controversial passenger. Commander Karol Bobko, pilot Don Williams and mission specialists Rhea Seddon, David Griggs and Jeffrey Hoffman were to have been accompanied by Frenchman Patrick Baudry and U.S. Senator Jake Garn. Baudry, who had served as back-up cosmonaut for the joint Soviet-French mission in August 1982, would perform medical experiments. One, called Echography, would measure blood flow in the body using sound waves.

Garn's presence was not universally welcomed. Some time before, President Reagan had announced that a teacher would be selected to fly as an ordinary citizen so it was with some surprise that Garn, chairman of the Senate committee that oversaw NASA's budget, was presented as the first non-specialist passenger. His duties included cooking and photography, as well as acting as a guinea pig for medical studies, but the political connection was widely viewed with cynicism. Certainly, the resulting publicity did NASA no good at all.

Hopes of a 51E mission died when engineers discovered that a design

Left:
Gardner celebrates the successful dual retrieval. The two satellites are barely visible at bottom right.

Military mission 51C crew with a launch pad escape vehicle. From left: Major Ellison Onizuka, Major Gary Payton, Commodore Tom Mattingly, Lt. Col. Loren Shriver and Lt. Col. James Buchli.

deficiency could allow outside interference to affect the TDRS-1 satellite launched by STS-6. This was serious since it not only routed NASA data but also military communications. TDRS-2 had to be grounded for modifications and 51E was scrapped. The satellite eventually made it off the ground aboard 51L.

The next up was 51D. Most of the 51E crew took over this flight but Baudry had to step down because Charles

Walker, again with the McDonnell-Douglas electrophoresis equipment, had higher priority.

It was to be a routine mission with no eye-catching space walks or satellite rescues but before it was over yet another comsat had been abandoned in orbit. Launch came on 13 April just a few seconds before the window closed at 14.00 GMT and the first day saw the $65-million Anik C1 comsat ejected and sent on its way. The same day, Senator Garn started two days of space sickness. This was just what the doctor ordered as Garn was carrying a set of microphones stuck to his midriff to record his internal rumblings.

The flight was going as planned until the Syncom 4-3 comsat release. It rolled smoothly out and the crew watched for a rod-like antenna to pop up as the first sign of life. Nothing happened. Seddon radioed down forlornly 'Houston, we are watching the Syncom and the omni antenna is still down'. Syncom was well and truly dead.

Engineers decided that the most likely culprit was the failure of an arming lever on the side and an army of astronauts and technicians sprang into action to come up with a plan to trip it. Otherwise, an $85-million craft would be added to the growing list of space junk.

The next day, Hoffman and Griggs put on their suits for the first unplanned space walk of the entire American space programme and within an hour had taped two devices made out of plastic report covers, bits of aluminium and sticky tape to the end of the arm. One had the appearance of a fly swatter while the other was a more rigid device like a lacrosse stick. The

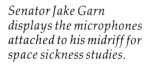

Senator Jake Garn displays the microphones attached to his midriff for space sickness studies.

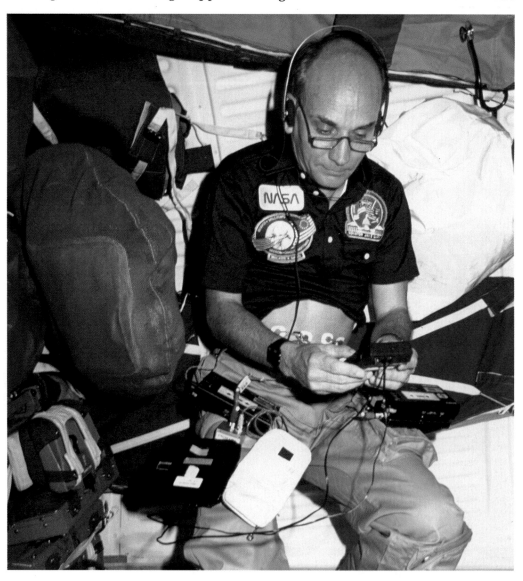

The arm and its two home-made devices wait for the arming lever to hove into view.

idea was to trip the lever firmly as it rotated past but if it got caught up the device would tear safely.

On day six, Bobko made the tricky moves to shift *Discovery* to within a few feet of Syncom and Seddon began flexing her arm. She jockeyed the robot appendage closer, missed the first time but then made good contact four times out of the next five. Syncom stubbornly remained dormant and later analysis of the pictures showed the lever had been fully extended all of the time anyway. It was clear that several failures must have occurred *inside* and no amount of tugging would have any effect.

Discovery headed towards Florida on 19 April after almost exactly seven days in space. Gusts of up to 17 mph (28 km/h) across the runway at the Kennedy Space Center forced the commander to apply differential braking to prevent his ship from drifting to the left and the right-hand wheel brakes locked up and the tyres burst. The ship was never in any danger but there could have been problems if a heavy cargo had been in the back – as *Challenger* was scheduled to carry on the next flight. The brakes had been troublesome throughout the flight programme and NASA decided that, until further notice, all future Shuttles would land in California where there was greater margin for error. This would increase the cost and time spent preparing each mission but the brake and crosswind problem had to be solved.

Meanwhile, engineers were looking at the possibility of a Syncom 4-3 rescue mission, with the $12-million cost footed by Hughes and the insurance companies. They hoped to use mission 51I in August somehow to bypass the failed electronics and get the comsat back on its feet.

Before then, however, the agency had three missions to complete, two of them highly complex Spacelab scientific flights.

97

The Future Is Not Free

The next Spacelab in line was in fact Spacelab 3 because Spacelab 2 had suffered serious equipment delays. 51B/Spacelab 3 made it into orbit on 29 April carrying a pressurized module similar to Spacelab 1 but this time it was occupied mainly by experiments to investigate biological responses and the processing of materials in weightlessness.

The Drop Dynamics Module was used by scientist Taylor Wang for levitating liquid droplets with sound waves to demonstrate the manufacture of very pure materials in the absence of contaminating containers. The equipment broke down early on but he was able to make repairs and achieve most of the original objectives.

Perhaps the most important experiment of the 15 aboard the $100-million flight grew a bright-red mercuric oxide crystal for 104 hours. On Earth, gravity creates structural faults in such a crystal but in space it could be grown larger and purer for use in X-ray and gamma-ray detectors. This test sample alone was worth several million dollars and will be used for years in research projects. Other crystals will follow and eventually the process should evolve into a commercial venture. For example, the Grumman aerospace company intends to manufacture gallium arsenide crystals for faster electronic chips by around 1988.

Taylor Wang manipulates a 0.6-inch (1.5-cm) diameter liquid droplet in the Drop Dynamics Module.

Above:
The Spacelab 3 crew of seven in the pressurized module. From left: Fred Gregory, Bob Overmyer, Don Lind, Norman Thagard, Bill Thornton, Taylor Wang and Lodewijk van den Berg.

Left:
Spacelab 3 found this spectacular aurora while flying between Australia and Antarctica.

Spacelab 3 also carried a set of $10-million animal cages for housing two squirrel monkeys and 24 rats as laboratory specimens. Unfortunately, the seals on the cages failed. Food and droppings floated out and the astronauts had to don surgical masks and fly around vacuuming up the debris! Commander Bob Overmyer made his feelings known to Houston in no uncertain terms as pieces whirled around *Challenger*'s cockpit.

The seven-day mission was declared to be a great success, as was the similar Spacelab D-1/61A flight the following November. This 22nd mission was paid for by West Germany to carry its own materials-processing experiments. The two German and one Dutch payload specialists, together with the five NASA astronauts, were directed in their scientific work from the Oberpfaffenhofen centre near Munich. As on Spacelab 1, they also used an instrumented helmet to record precise eye and head movements for spacesickness studies but this time the wearer could be strapped into a seat and accelerated down a track running the length of the pressurized module.

Of the 75 experiments performed by the crew of eight, only one failed. The West German space agency claimed that 80% of the research objectives had been met when *Challenger* landed on 6 November. Its next mission was to be 51L the following January.

The West German flight was the fourth Spacelab mission, having been preceded by 51F, which carried Spacelab 2 aloft. This Spacelab was rather different to the others in that it carried mainly astronomical telescopes mounted on three pallets in the payload bay. There was no pressurized module and the crew of seven had to live and work in *Challenger*'s cramped cabin for a full week.

The 50th American manned spaceflight, with mission specialist Karl Henize at 58 the oldest space voyager so far, attempted to leave Pad 39A on 12 July 1985 but after the main engines had been burning for three seconds, computers detected a sticking fuel valve on engine 2 and aborted the flight before the solid boosters were ignited. The crew stayed aboard for 45 minutes while the external tank was drained and they were able to walk off along the access arm.

The next attempt came on 29 July and there appeared to be no difficulties until, 5 minutes 43 seconds into the flight, the centre engine was closed down by the onboard computers. It later turned out that a faulty sensor was to blame and the engine could have continued to fire.

The spacecraft communicator in Houston radioed up 'Abort ATO, Abort ATO', meaning that *Challenger* should Abort-to-Orbit on the remaining two engines. Commander Gordon Fullerton clicked a dial to the ATO setting and ordered the computer to carry out the abort program. If shutdown had come 33 seconds earlier, *Challenger* would have had just enough speed to limp across the Atlantic to land at Zaragosa in Spain. As it was, the spaceplane dumped 4400lb (2000 kg) of OMS propellants from the pods to lighten its load and reached a 196-mile (315-km) orbit instead of the planned 236 miles (380 km).

In orbit, the crew settled down to their astronomical observations. The main item in the cargo bay was Europe's $60-million Instrument Pointing System with its complement of four telescopes for Sun observations. The Shuttle is an unsteady platform by scientific standards and the IPS is designed to hold up to 2 tons/tonnes of instruments rock solid. There were problems getting its controlling software working but eventually it came

Opposite:
An unusual perspective on the launch of the Spacelab 2 mission.

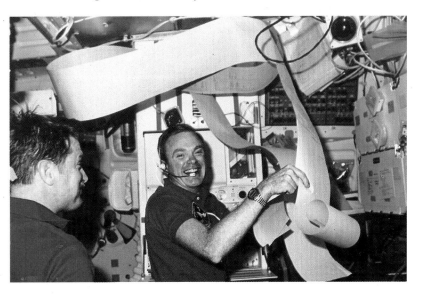

Tony England struggles with a stream of teleprinter roll holding instructions from ground controllers. Roy Bridges looks on.

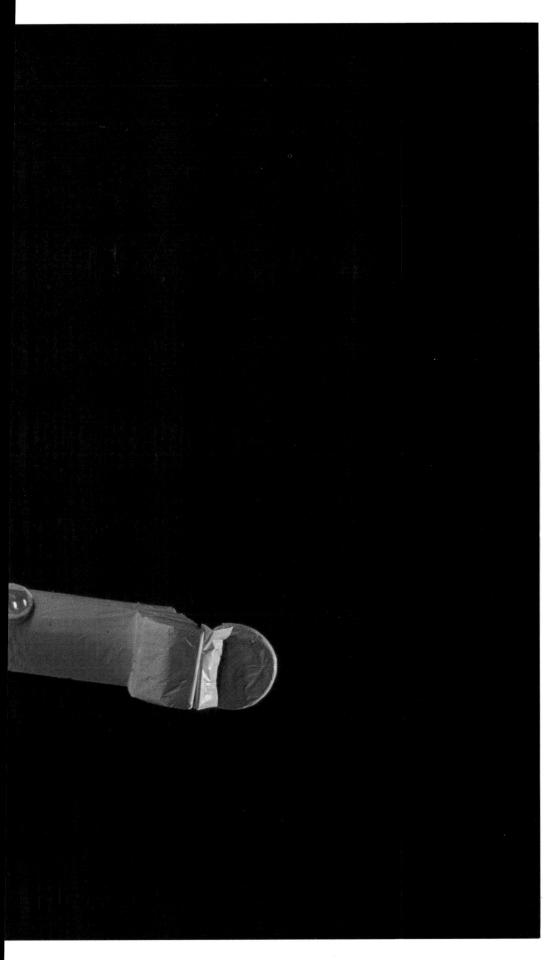

The European Instrument Pointing System (IPS) can direct several tons/tonnes of astronomical instruments very accurately.

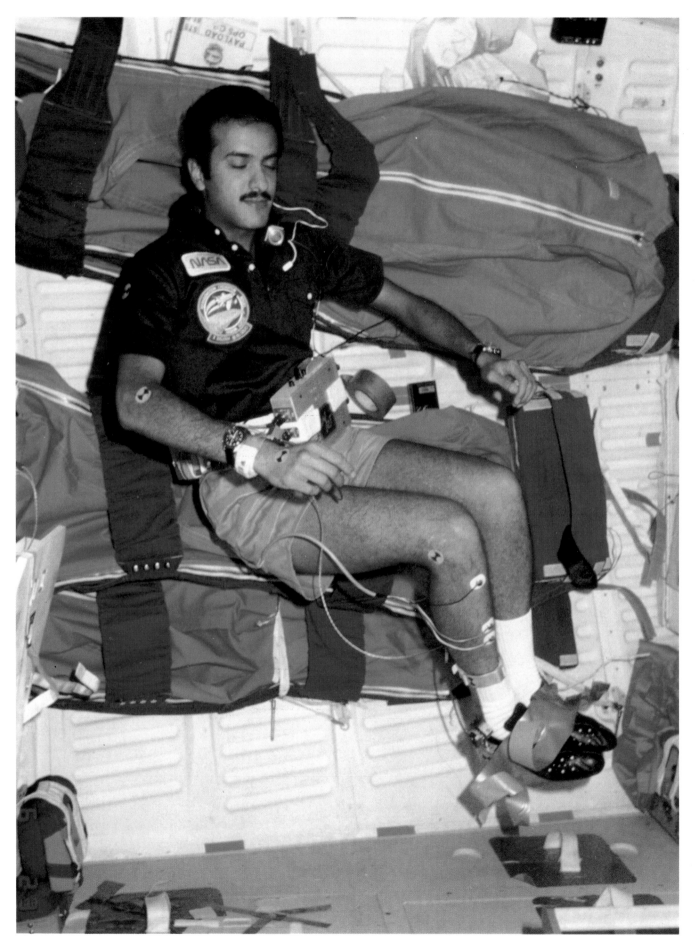

through with flying colours. It should have also seen service aboard 61E in March 1986 to observe Halley's Comet at the same time as the probes flew by but the 51L explosion intervened. It was a serious loss of scientific data.

The Sun was extensively observed at different wavelengths during Spacelab 2 and a U.K. X-ray telescope looked at emissions from distant galaxies. An infrared telescope found a background glow that will be troublesome for the future – it apparently came from water vapour surrounding *Challenger*. Another experiment observed cosmic rays arriving from deep space and instruments recorded how the spacecraft disturbed the plasma (charged atoms and particles) surrounding the Earth.

When Gordon Fullerton brought *Challenger* to a halt on the runway in California on 6 August, NASA was well satisfied with the Shuttle's progress. *Challenger* was towed away to be prepared for the West German Spacelab D-1 mission and attention turned to *Discovery*.

An International Flavour

Slotted between the two highly successful Spacelab 2 and 3 missions was the international 51G flight. It achieved nothing dramatic (the release of comsats was by now so routine that it attracted little interest) but NASA officials rated it as the most successful thus far.

Media attention focused on two aspects: a rather minor Star Wars experiment and the inclusion of an Arab prince and a Frenchman in the crew. The American team of Dan Brandenstein (commander), John Creighton (pilot), John Fabian, Steven Nagel and Shannon Lucid (mission specialists) were joined by Patrick Baudry and Prince Sultan Salman Al-Saud for launch on 17 June 1985.

Baudry's main task was to perform medical tests updated from the French-Soviet flight, measuring blood flow and recording body posture and movements. The Prince, on the other hand, had little to do except serve as a test subject for Baudry and take photo-

Patrick Baudry conducts his own posture tests to see how the human body copes with weightlessness.

graphs of the Arab countries to help in the search for oil. A nephew of King Fahd of Saudi Arabia, he was a TV advertising executive and had no responsibility for the Arabsat 1B comsat sitting in the payload bay. That was left to the professional astronauts. U.S. officials were not slow to point out that the first Arabsat launch aboard Europe's Ariane had cost $23.75 million, while the Shuttle produced a bill for only $20.5 million *and* could carry a passenger. The commercial benefit stemming from NASA's offer to fly representatives with major payloads had been borne out.

Apart from Arabsat, Mexico's Morelos 1 and the U.S. commercial Telstar

Opposite:
Prince Sultan Salman Al-Saud takes part in Baudry's posture experiment.

105

Right:
William Fisher (Anna Fisher's husband) is anchored to the side of the cargo bay to work on the captured Syncom.

3D comsats were also successfully. On day four, Shannon Lucid used the arm to release Spartan, a new and relatively cheap type of retrievable satellite. Spartan was a simple, box-like 1-ton/tonne structure carrying, in this case, an X-ray telescope to observe radiation from the Milky Way and other galaxies.

The Star Wars test involved bouncing a laser beam fired up from Hawaii off a set of mirrors attached to the main hatch window to demonstrate laser tracking. Unfortunately, an embarrassing slip caused *Discovery* to be pointing completely in the wrong direction and a second attempt had to be made before everyone got it right.

After slightly more than seven days aloft, *Discovery* landed at Edwards on 24 June to end the 18th mission. For mission specialist John Fabian, it was his second and last landing. He had already been selected to fly aboard the important 61G in May 1986 to release the Galileo Jupiter probe but he announced his retirement from the astronaut corps in September 1985, the first crew member known to have refused a flight. He said that preparing for missions left him too little time with his family: 'We get the benefit of going into space. The families don't.' His resignation left NASA's astronaut team standing at 102.

Space Rescue – Again

While *Challenger* carried the two Spacelabs and *Discovery* flew 51G, the dormant $85-million Syncom 4-3 was coasting silently around the Earth instead of acting as a U.S. Navy communications relay 22,300 miles (35,900 km) above the equator. Its failure following release from 51D the previous April precipitated a crash programme to repair it in orbit: NASA was eager to demonstrate the Shuttle's flexibility once again and Hughes, the builders, wanted to save face and avoid financial penalties from the waiting military users.

The 20th Shuttle mission, and *Discovery*'s sixth, finally left the ground on 27 August 1985 after two scrubbed attempts. Bad weather had forced a postponement on the 24th and the

'Ox' van Hoften pushed on Syncom several times to send it spinning away.

failure of computer 5 cost another 48 hours. Even so, the 4,517,225-lb (2,049,000-kg) vehicle only just made it through a break in the thick cloud cover blanketing the Cape.

Discovery carried three fare-paying passengers for release during the first three days. Syncom 4-4, a twin of the repair target, was to become the third link in the U.S. Navy network. Aussat 1 was the first of three national Australian comsats and ASC-1 belonged to a U.S. commercial operator. Aussat should have been released on day two but the robot arm hit its cradle's protective sunshield and it had to be pushed out during the fifth Earth circuit. Because of the slight damage, the sunshield was locked open and the satellite would have been exposed to unpleasant temperature variations if it had stayed put. ASC-1 followed it on the eighth orbit, the first time a crew had coped with two major releases on the same day. Day three saw Syncom 4-4 safely on its way.

Discovery's path was carefully controlled by commander Joe Engle and pilot Dick Covey to catch up with Syncom 4-3 during day five (31 August). As they moved their ship to within a few feet, William Fisher and James van Hoften waited in the payload bay to begin what was to become, at 7 hours 8 minutes, the longest American space walk in Earth orbit. Mike Lounge stood by to operate the arm although he had an additional difficulty to cope with when the crew found control of the elbow joint had failed on the first day. It did not seem to be linked to the sunshield incident and no one blamed the crew, but it did mean Lounge had to flip switches on his rear-facing control panel to move the elbow one direction at a time.

As Syncom came within range, van Hoften reached out from his platform on the end of the arm above the bay, while Fisher stood attached to the starboard wall. Syncom was spinning very slowly and van Hoften easily closed the arming lever on its side, reversing the work Rhea Seddon had done the previous April with her 'flyswat'. The satellite's internal timer could not now begin ticking away and allow the motor to flare into life.

Van Hoften and Fisher snapped handling bars into place for *Discovery*'s arm to grab hold of the satellite and free them for the repair work. Fisher plugged in an electronics box for ground controllers to send commands into Syncom; a power device was attached to the omni antenna to raise it.

Following a night's rest, van Hoften gave Syncom 'a heck of a push' to send it spinning away. Controllers kept it in low orbit until 27 October to give the solid-propellant boost motor time to warm up in the sunlight. It ignited cleanly and Syncom ended up operating successfully in geostationary orbit.

Joe Engle commented, after the landing in California on 3 September, 'It was one of the most fantastic things I've ever been involved with. Five months ago we had no idea we'd be doing this, and four months ago we were not sure it would really happen.'

Ironically, the Syncom 4-4 satellite released earlier in the mission failed on 6 September after reaching geostationary orbit. It had been modified of course to prevent problems similar to its predecessor's but it appears that the cable linking the transmitter and antennae failed. Already in high orbit, it is beyond the Shuttle's reach and effectively an $85-million write-off.

But this was not NASA's concern. The 20th Shuttle flight had gone amazingly well and it was left to newcomer *Atlantis* to declare the system of age with flight 21.

A Secret Mission

When *Atlantis* joined its three stablemates in April 1985 to complete the fleet, it carried only 21,801 thermal tiles as lightweight blanket material had once again been used in less sensitive areas.

As *Atlantis* left Rockwell in Cali-

fornia, the company laid off workers to bring the production line to an end. It still appeared to NASA that a fifth orbiter was unnecessary but, even so, they allowed an impressive, $110-million set of spares to build up: wings, tail, fuselage, crew module and payload-bay doors. Extensive damage to an existing orbiter would require a big rebuilding job and there were possibly enough spares in storage to form the basis of another spaceplane.

On 12 September 1985 the three main engines grouped at the rear of *Atlantis* roared into life for a mere 22 seconds as engineers checked the new systems.

The complete crew for 51J was announced only shortly before the test since this was to be the second all-military flight. The NASA contingent of Col. Karol Bobko (commander), Lt. Col. Ronald Grabe (pilot) and mission specialists Col. Bob Stewart and Major David Hilmers had been known for

The DSCS-3 comsats will be an integral part of the U.S. military communications network until at least the end of this century.

some time, but the addition of Major William Pailes was held back until only a month before launch was due.

The exact time of the launch itself was kept under wraps – it would fall within a three-hour period on 3 October 1985 – and the nature of *Atlantis'* cargo was kept secret.

The flight appears to have gone exactly according to plan but few details were released. It was noted that *Atlantis* broke the Shuttle altitude record by reaching 320 miles (515 km), half of the maximum theoretical height. Landing in California came on 7 October, keeping to the pattern for short military missions.

Although the U.S. Air Force refused to disclose which satellites were released, it was generally acknowledged they were two DSCS-3 military comsats borne by a single Inertial Upper Stage. Each should survive for ten years relaying U.S. military communications around the world.

Being a military mission, 51J did not celebrate the 21st trip into orbit in a spectacular manner but it was perhaps appropriate that newcomer *Atlantis* received the honour. In those 21 missions, lasting for a total 132½ days, 78 individuals had flown, 18 of them twice, with Karol Bobko making it three times and Bob Crippen four. Six women had flown – Sally Ride the only one to do so twice – together with four foreigners, two military engineers, seven non-astronaut payload specialists and one 'Congressional Observer', Jake Garn. All in all, the flight experience was equivalent to one person spending two years in space. All 35 of the Group 8 astronauts had gone up. They were selected in 1978 especially for the Shuttle programme and their short wait was in sharp contrast to the dedication of Joe Engle, who was part of the 1966 intake. He lost out in Apollo and Skylab and finally reached orbit in November 1981 aboard STS-2 – 15½ years later!

The Group 9 astronauts, selected in 1980, made 11 appearances in the first 21 trips and they will become increasingly active. Similarly, more Group 8 people will receive command positions. Fred Hauck was the first in 1984 when he commanded 51A.

Towards The Space Station

When the U.S. Space Station is established in orbit in the 1990s, a large part of it will be an open framework for attaching equipment, modules and platforms. Nothing like it has ever been assembled in space and NASA's engineers need to test the efficiency of various methods of construction under realistic conditions. Mission 61B, the 23rd in the series, allowed mission specialists Jerry Ross and Sherwood Spring to spend 13 hours erecting and dismantling two sets of frameworks in the payload bay, all the while being filmed by stereo cameras to produce accurate time-and-motion studies for future reference.

Atlantis leapt into Florida's night sky – only the second Shuttle launch in darkness – on 26 November 1985 with a full payload bay. The first satellite released was Mexico's Morelos 2 comsat, watched by Mexican telecommunications specialist Rudolfo Neri Vela. The eighth foreigner to fly, he was chosen from 800 applicants, although the devastation caused by the recent Mexican earthquake meant that the satellite was not yet required for the country's expanding communications network. It was cheaper, though, to let it take two years to drift into final orbital position than keep it in storage and pay NASA's increased launch charges later on.

Second out was Australia's Aussat 2 national comsat, for which they paid NASA $9.5 million in launch costs. The third, and most powerful domestic comsat ever orbited, was the commercial Satcom K-2. The loss of several expensive comsats over the previous 20 months had cost insurers around $600 million and they demanded a premium of 30% on this particular satellite. The RCA company decided to take a $75 million gamble and fly without cover.

Aboard for his third flight was Charles Walker, the engineer accompanying McDonnell-Douglas' electrophoresis unit, hoping that he could process enough of the anaemia-combating erythropoietin hormone for animal and clinical trials. A second commercial venture was the 3M Cor-

Sherwood Spring moves the ACCESS tower around high above the Gulf of Mexico.

poration's unit to mix organic solutions to produce pure crystals with better optical and electronics properties than those grown on Earth. One aim is to manufacture more efficient fibre optics.

Satellite deployments and materials processing had now become routine, so it was natural that most attention was focused on the two space walks. Ross and Spring left the airlock on 29 November to use the ACCESS (Assembly Concept for Construction of Erectable Space Structures) kit. Standing in foot restraints, they built up a triangular framework 45 ft (13.7 m) long out of its 93 tubular aluminium struts and 33 connecting nodes using only their hands. Allocated two hours for the task, they assembled the 10 bays in only 58 minutes and then took them apart.

A second, very different, kit now came into play: EASE (Experimental Assembly of Structures in EVA). The two astronauts floated free this time as they created an inverted-pyramid framework out of the six 12-ft (3.66-m), 64-lb (29-kg) beams no less than eight times. Fatigue will be an important factor in building future space stations and Spring did complain of aching hands.

Following a rest day, the pair emerged again. This time they erected nine of the ACCESS bays and then

Spring and Ross assemble the EASE pyramid.

Ross climbed on the end of the robot arm for Mary Cleave to hold him at the top end while he added the tenth. He also clipped a length of rope along the frame to simulate electrical cabling.

Spring released the base and his partner man-handled the whole length to see how easy it was to control and whether he could slot it back in place. He had no problems. Spring took a turn on the arm platform and disassembled bay 10 before simulating repair work on bay 8 and handling the freed structure himself.

The astronauts also built the EASE pyramid again but this time used the arm platform instead of floating around.

As the 6½-hour walk ended, the tired but contented mission specialists

An unusual low-angle view of Columbia *shortly after ignition for mission 61C on 12 January 1986.*

had made a significant contribution to space engineering, providing enough information to keep designers occupied for years. Simulations in large water tanks can now be carried out with greater realism.

Commander Brewster Shaw and pilot Bryan O'Connor steered *Atlantis* down in California on 3 December to clear the way for *Columbia*'s first mission since STS-9 in December 1983. While the other three orbiters shouldered 14 missions, *Columbia* had spent 18 months at Rockwell being converted from its test configuration. The ejection seats were removed, for example. NASA took the opportunity to add a new type of nose cone and a camera pod at the tip of the tail for recording aerodynamic and thermal data. The high-altitude, hypersonic region of flight is difficult to simulate in wind tunnels and *Columbia* can now collect air-pressure data from 14 holes in its nose cap and measure the temperatures of the upper wing and fuselage surfaces from the tail pod. This information will help in designing future hypersonic passenger aircraft and follow-on Shuttle vehicles.

Almost 300 wing-load sensors were added to learn more about the Shuttle's response to launch stresses as *Columbia* undertook the most demanding flight profile in the programme. Wind-tunnel tests had proved inaccurate and the Shuttles so far had been falling short of their design payload capacity. The sensors would help engineers to understand the problem.

The last Syncom comsat had been dropped from 61C's manifest as RCA's designers strove to avoid a repeat of Syncom 4-4's fate following release from 51I the previous August. Gregory Jarvis, the RCA specialist who should have accompanied Syncom, was replaced in the crew by politician Bill Nelson, chairman of the House Space Science and Applications subcommittee. Jarvis was found a seat next to Christa McAuliffe on the ill-fated 51L in January instead.

RCA still had their smaller Satcom

Columbia *was accompanied in orbit for a while by this piece of thermal tile.*

Congressional Observer Bill Nelson participates in a medical test aboard Columbia.

George Nelson involved in photographing Halley's Comet as part of CHAMP (Comet Halley Active Monitoring Program).

K-2 aboard and this time they had secured insurance coverage of $80 million at a premium of $16 million. Their representative was Robert Cenker, who would test a new infrared detection system to track U.S.A.F. aircraft as part of the Star Wars research programme. The U.S.A.F. Teal Ruby satellite due on the first Shuttle launch from Vandenberg Air Force Base in summer 1986 carries a similar detector.

Columbia's cargo bay also held Materials Science Laboratory-2 with three experiments to test levitation methods in zero gravity and the production of magnetic composite materials. The first GAS Bridge held a dozen Getaway Special canisters in an attempt to clear the growing backlog. One was designed to see how weightlessness affected the crystalline structure of cancer-fighting substances. Bill Nelson also intended to supervise the growth of protein crystals in the cabin for medical research.

Hitchhiker made its debut, a more sophisticated GAS capable of holding larger experiments with data links to the orbiter. GAS canisters are self-contained and their success can be gauged only after landing. Hitchhiker's main customer this time was NASA equipment to test cooling methods for the Space Station.

The orbital portion of the mission was unremarkable except that two of the three MSL-2 experiments failed and astronomer Steven Hawley had to cope with a broken light intensifier when he tried to photograph Halley's Comet. 61C was mainly notable for the launch and landing sagas. Lift-off was planned for 18 December 1985 but four scrubs pushed it well into the New Year. On 19 December it had got to within T-14 seconds when the launch computer detected a fault in the right-hand solid booster's hydraulic power unit, used to swivel the nozzle for steering. Bad weather also caused delays.

Columbia eventually climbed into the pre-dawn skies on 12 January 1986 but NASA was eager to get the flight over. With 14 other missions and three orbiters in line over the year, turn-around time on the ground was short. It was helpful that 61C would make the first landing back in Florida since 51D in April 1985, thus saving a week otherwise spent transporting it back from California. The difficulty was that 61E in March had to get off on the 6th or soon after, and the Galileo Jupiter and Ulysses Sun probes also had tight windows, in May.

The scheduling problem was becoming so acute that NASA decided to lop a day off 61C and bring it back down at KSC on 16 January. Unfortunately, the weather disagreed and not only was

---

SPACE SHUTTLE MISSIONS TO JANUARY 1986

Columbia:	1, 2, 3, 4, 5, 9, 61C	7 missions
Challenger:	6, 7, 8, 41B, 41C, 41G, 51B, 51F, 61A, 51L	10 missions
Discovery:	41D, 51A, 51C, 51D, 51G, 51I	6 missions
Atlantis:	51J, 61B	2 missions

In the 24 completed missions, 91 individuals flew, 25 of them twice, 3 three times and only one (Crippen) four. The total include 8 women, 8 foreigners, 2 Congressional observers and 2 U.S.A.F. specialists.

SHUTTLE'S ORIGINAL 1986 LAUNCH SCHEDULE

Vehicle	Mission	Launch	Payload
Columbia	61C	12 Jan	RCA Satcom K-1 comsat
Challenger	51L	28 Jan	TDRS-2 comsat
Columbia	61E	6 Mar	Astro-1 Halley's Comet telescope
Challenger	61F	15 May	Ulysses Sun probe
Atlantis	61G	20 May	Galileo Jupiter probe
Columbia	61H	24 Jun	3 comsats + first British astronaut
Discovery	62A	July	Military; first from California
Challenger	61M	22 July	TDRS-3 comsat + drugs processing
Atlantis	61K	18 Aug	Earth Observing Mission
Columbia	61N	4 Sep	Military mission
Challenger	61I	27 Sep	Indian comsat + satellite retrieval
Discovery	62B	29 Sep	Military; second from California
Atlantis	61J	27 Oct	Hubble Space Telescope
Columbia	61L	6 Nov	Two comsats + materials processing
Challenger	71B	6 Dec	Military mission

Overleaf: Columbia *was the first orbiter to be used and will continue to see service for many years. Here it is releasing a satellite powered by the new upper stage built by the Orbital Sciences Corporation.*

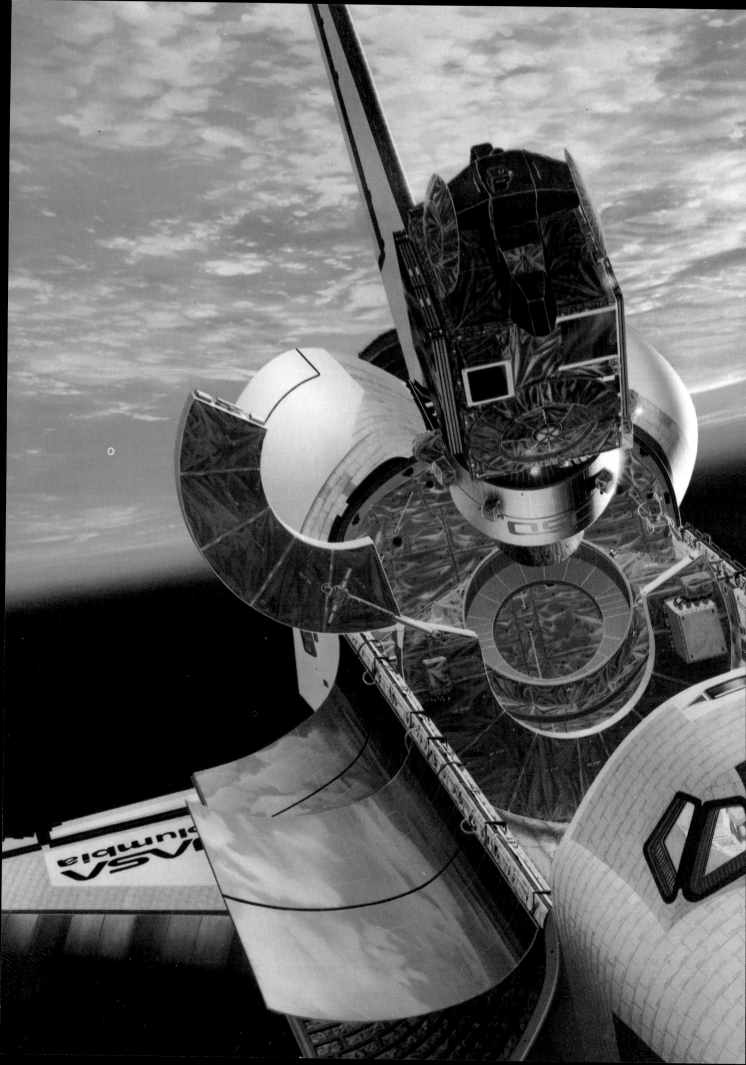

RCA engineer Robert Cenker displays his technical skill and repairs some equipment aboard Columbia.

Pilot Charles Bolden prepares for re-entry. Mounted in front of his window is a Head Up Display which allows him to read flight information without taking his eyes away from the outside view.

the landing delayed to the 18th but it also had to be switched to California. Robert Gibson and Charles Bolden guided *Columbia* down for only the second night landing and an hour later the smiling crew of seven disembarked. They were the last crew to do so for some time.

Back in Florida, technicians were preparing *Challenger* for its tenth launch while its astronauts put the finishing touches to their training.

'Star Voyagers'

Shuttle mission 51L, the heaviest launched so far, was intended to be a milestone: it brought the programme to the quarter-century mark and would take the total of world space travellers above 200. When the crew of seven climbed aboard on the morning of Tuesday, 28 January 1986, they had several days of frustration behind them. The previous day's attempt had been called off when technicians failed in their struggle to remove a handle after closing the hatch.

This time, though, there were no postponements. In the commander's left-hand cockpit seat was Dick Scobee, a 47-year old Vietnam veteran and already with 41C in April 1984 under his belt. To his right was new man Michael Smith, another Vietnam pilot. Behind them in the centre seat, acting as flight engineer to help out with checklists during launch, was Judith Resnik. On mission 41D she had become the second woman astronaut and she was included on this trip because of her expertise with the robot arm. It would be her job to release the Spartan-Halley astronomical satellite, allowing it to fly separately for 45 hours making unique ultraviolet observations of Halley's Comet. Ronald McNair, a veteran of 41B and now sitting next to the hatch downstairs, was responsible for operating the satellite.

Challenger's third mission specialist was Ellison Onizuka, sitting behind pilot Smith. He was aboard to look after the major payload: NASA's large $100-million TDRS-2 comsat with its attached Inertial Upper Stage. Onizuka had already dealt with an IUS-borne satellite during the 51C military mission in January 1985 and expected to wave goodbye to his second ten hours after launch.

The other two crew members were not professional astronauts. Gregory Jarvis was an RCA company engineer flying with a set of containers to see how liquid sloshes around in weightlessness, aiming to improve the design of satellite propellant tanks. Christa McAuliffe was aboard as a result of NASA's policy of opening up the space programme to ordinary people. A 37-year old social sciences teacher at Concord High School in New Hampshire, she was one of 10,690 American educators who had applied for the Teacher in Space Project by the closing date of 1 February 1985. As one of the 114 semi-finalists and then the ten finalists, she underwent a week of psychological and physical tests, including trips aloft in NASA's 'Vomit Comet' aircraft to see how she reacted to periods of weightlessness. The teachers also underwent written tests, having to explain why they wanted to go into space and how they planned to use their experience for educational purposes. McCauliffe's selection was announced in the summer of 1985, along with that of her back-up, Barbara Morgan, in time for eight weeks' basic training.

McCauliffe was keeping a detailed diary and was due to conduct two live lessons from orbit on day six. 'The Ultimate Field Trip' would be a guided tour around the Shuttle, while 'Where We've Been, Where We're Going And Why' was a history of the space programme. She would also make an educational video illustrating aspects of weightlessness for distribution to schools. The pair of non-astronauts sat together in front of the airlock with no view out.

The bitter January weather had again threatened the launch but at 11.38 a.m. local time (16.38 GMT) *Challenger* lit its main engines and after the computers had declared them to be healthy the solid boosters were ignited and launch pad 39B released its grasp.

Challenger arced out over the Atlantic, rolling over on to its back as planned. The boosters were midway through their burn when the pilots

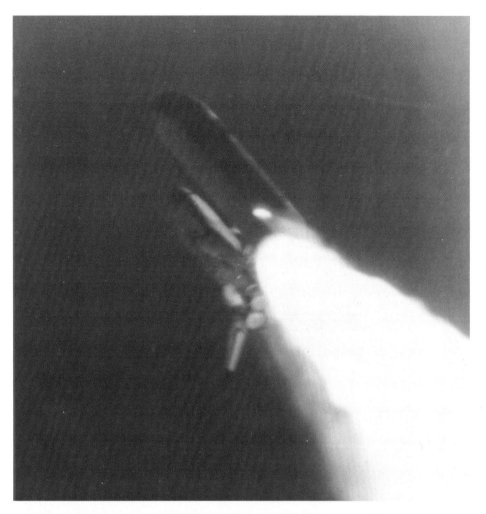

This picture of a released external tank highlights the liquid oxygen line that is suspected to have ruptured on Challenger *after the tank base had been engulfed in flames from the right-hand booster. The liquid hydrogen line emerges directly from the base dome.*

received the go-ahead from astronaut Dick Covey in Houston to turn the main engines up to full power. Maximum thrust is avoided early on to prevent the Shuttle being pushed too hard in the thicker layers of the atmosphere. The period of greatest aerodynamic pressure had just passed as Challenger accelerated to about 2000 mph (3200 km/h) some 10 miles (16 km) high. The last words heard from the vehicle were Scobee's 'Roger, go at throttle up'.

At this stage, 72 seconds after launch, flames appeared around the base of the external tank. Suddenly, the tank exploded and *Challenger* was engulfed. The two boosters somehow separated and continued skywards before the range safety officer destroyed them 30 seconds later when one threatened to return to Florida.

Of *Challenger*, the only sign was a shower of debris splashing into the cold Atlantic waters.

The world was stunned. The first American in-flight fatalities in 25 years brought the programme to a halt as NASA set up review boards to sift through the masses of evidence. It was vitally important to recover all the wreckage possible and large chunks of *Challenger* were soon being landed at Port Canaveral. Tracking film quickly suggested that flames had initially appeared from the side of the left-hand booster. Footage released a few days later showed the other side: it appeared there was a rupture in the

 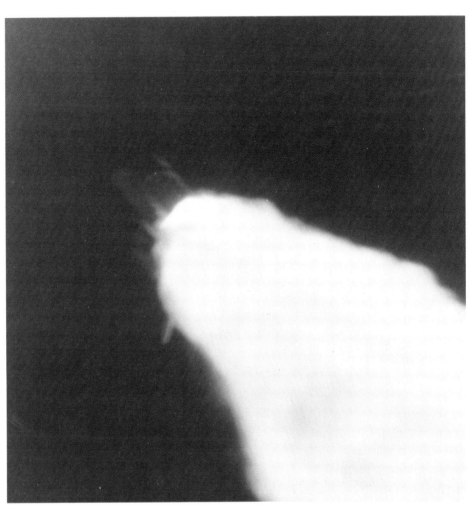

right-hand booster's steel casing, with flames playing on the external tank for some 13 seconds before the explosion. Telemetry analysis showed a 5% drop in SRB pressure and a reduction in oxygen flowing to the main engines, consistent with a breached booster and a damaged oxygen pipe running along the tank under *Challenger*'s belly. The wisdom of using segmented, reusable boosters to carry people became a leading topic for President Reagan's investigating commission, due to report by the end of May.

The American space programme has suffered a severe setback and two likely consequences are that non-essential personnel will not fly again for some time and that the Pentagon will move back towards the old, expendable rockets. Military space planners had already said that national security was too important to entrust solely to the Shuttle and had ordered ten new Titan 3 rockets to begin launching spy satellites in 1988. Building a new orbiter to replace *Challenger* would be very expensive and take several years. Military payloads have first call on the three remaining spacecraft, which could fly 12 to 15 missions a year, so commercial and, particularly, scientific cargo will inevitably suffer.

At the Johnson Space Center in a memorial service on the Friday following the tragedy, President Reagan said that the bereaved families had all expressed their desire for the programme to continue. It was what *Challenger*'s crew – the 'Star Voyagers' he called them – would have wanted.

'The Earth is the cradle of the mind, but one cannot live forever in a cradle'
Konstantin Tsiolkovsky space pioneer, 1899

Far left:
Challenger *58 seconds after launch, showing an unusual exhaust plume on the right-hand booster.*

Centre:
Sixty seconds after launch and the rupture in the booster is already spraying flame on to the external tank.

Above:
The spacecraft a split second before explosion. This sequence of three pictures was captured by a 70mm-camera north of the launch site.

Overleaf:
Shuttle mission 51L erupts into a ball of flame. The two boosters continue skywards on their plumes of smoke.

Shuttle Flight Summary

STS-1
Launched 12 April 1981 with John Young and Robert Crippen aboard *Columbia* for 2-day 6-hour test. Landed at Edwards Air Force Base, California.

STS-2
Columbia launched 12 November 1981 with Joe Engle and Richard Truly. Robot arm tested successfully. Five-day flight cut to 2 days because one of three electricity-producing fuel cells failed.

STS-3
Columbia launched 22 March 1982 with Jack Lousma and Charles Fullerton on 8-day test mission. Landed at Northrup Strip in New Mexico because of poor conditions in California.

STS-4
Columbia in last test flight, launched on 27 June 1982 for 7-day mission with Tom Mattingly and Henry Hartsfield. Two boosters were lost.

STS-5
First operational flight; *Columbia* carried Anik and SBS commercial comsats. Crew of four for first time: Vance Brand, Robert Overmyer and mission specialists Joe Allen and William Lenoir. First Shuttle space walk cancelled because of suit problems. Launched 16 November 1982; landed in California 22 November.

STS-6
First flight of *Challenger*. Launch 4 April 1983 with Paul Weitz, Karol Bobko and mission specialists Don Peterson and Story Musgrave. Large NASA TDRS comsat in trouble after Inertial Upper Stage fails. Peterson and Musgrave make first Shuttle space walk to test suits. Landed 9 April in California.

STS-7
Challenger carries first crew of five and first American woman: Bob Crippen, Fred Hauck and mission specialists John Fabian, Sally Ride and Norman Thagard. Released Anik and Palapa comsats. Test pallet satellite released and recovered with robot arm. Dr. Thagard worked exclusively on space-sickness studies. Launched 18 June 1983, landed 24 June in California.

STS-8
Challenger launched 30 August 1983 with Richard Truly, Dan Brandenstein and mission specialists Dale Gardner, Guion Bluford (first U.S. black astronaut) and Bill Thornton. First night launch and landing (5 September). Released Indian Insat 1B comms/weather satellite. Tested robot arm with large Payload Flight Test Article.

STS-9
Columbia carried first crew of six for Spacelab 1 mission. Pilots: John Young, Brewster Shaw; mission specialists Owen Garriott, Bob Parker; payload specialists Byron Lichtenberg, Ulf Merbold (West German). Ten-day test of space laboratory, 71 experiments. Launched 28 November 1983, landed 18 December in California.

41B
Challenger launched 3 February 1984 with Vance Brand, Bob Gibson and mission specialists Ron McNair, Bob Stewart and Bruce McCandless. Palapa and Westar comsats both lost when their motors failed; recovered in mission 51A November 1984. First test of Manned Maneuvering Unit backpack – McCandless flies free of Shuttle and becomes human satellite. Landed 11 February, the first to land at Kennedy Space Center.

41C
Challenger launched 6 April 1984 carrying Bob Crippen, Francis Scobee and mission specialists Terry Hart, James van Hoften and George Nelson. Released Long Duration Exposure Facility with 57 experiments to be picked up in 1985. Solar Max solar astronomical satellite rescued and Nelson and van Hoften repair it in payload bay for release. Landed 13 April in California.

41D
Discovery's first flight. Launched 29 August 1984 with pilots Henry Hartsfield, Michael Coats; mission specialists Judith Resnik, Steven Hawley, Richard Mullane; and payload specialist Charles Walker. Released three comsats: SBS, Telstar and Syncom. Non-astronaut Walker operated electrophoresis equipment for drugs research. Landed 5 September in California.

41G
Challenger launched 5 October 1984 with Bob Crippen (4th Shuttle flight), Jon McBride, mission specialists Kathryn Sullivan, Sally Ride, David Leestma; payload specialists Marc Garneau, Paul Scully-Power (oceanographer). Sullivan makes first American female space walk, tests refuelling of satellite in space. Release Earth Radiation Budget Satellite. Landed 13 October at K.S.C.

51A
Discovery launched 8 November 1984 with Fred Hauck, David Walker; mission specialists Anna Fisher, Dale Gardner, Joe Allen. Allen and Gardner rescue Westar and Palapa satellites stranded after mission 41B; brought back to Earth. Landed 16 November in California.

51C
Discovery launched 24 January 1985 with Tom Mattingly, Loren Shriver; mission specialists James Buchli, Ellison Onizuka and Gary Payton (U.S. Air Force astronaut). They release a military communications spy satellite and land at the Kennedy Space Center on 27 January.

51D
Discovery launched 13 April 1985 with Karol Bobko, Donald Williams; mission specialists Rhea Seddon, Jeffrey Hoffman, David Griggs; payload specialists Charles Walker and Congressional observer Jake Garn. Syncom, second of two comsats, is left in low orbit after attempts to revive it fail. Hoffman and Griggs make unplanned space walk. See mission 51I. Landed 19 April at Kennedy Space Center.

51B
Challenger launched 29 April 1985 with Bob Overmyer, Fred Gregory; mission specialists Don Lind, Norman Thagard, Bill Thornton; payload specialists Lodewijk van den Berg, Taylor Wang. Spacelab 3 mission carrying multitude of materials processing and life-sciences experiments in Spacelab module. Landed 6 May in California.

51G
Discovery launched 17 June 1985 with Dan Brandenstein, John Creighton; mission specialists Shannon Lucid, Steven Nagel, John Fabian; payload specialists Patrick Baudry, Prince Sultan Salman Al-Saud. Three comsats are released, Spartan observes X-rays and laser beam bounces off mirror in Star Wars test. Landed 24 June in California.

51F
Challenger launched 29 July 1985 with Charles Fullerton, Roy Bridges; mission specialists Story Musgrave, Tony England, Karl Henize; payload specialists Loren Acton, John-David Bartoe. Spacelab 2 mission devoted to astronomy using IPS and other instruments. Landed 6 August in California.

51I
Discovery launched 27 August 1985 with Joe Engle, Richard Covey; mission specialists James van Hoften, Mike Lounge, Bill Fisher. Three comsats released, van Hoften and Fisher space walk to repair Syncom satellite stranded after 51D. Landed 3 September in California.

51J
Atlantis (first mission) launched 3 October 1985 with Karol Bobko, Ronald Grabe; mission specialists Robert Stewart, David Hilmers, William Pailes (U.S.A.F. astronaut). Two DSCS-3 military comsats are released. Landed 7 October in California.

61A
Challenger launched 30 October 1985 with record eight-man crew of Henry Hartsfield, Steven Nagel; mission specialists James Buchli, Guion Bluford, Bonnie Dunbar; payload specialists Reinhard Furrer (W. Germany), Ernst Messerschmid (W. Germany), Wubbo Ockels (Holland). Spacelab D-1 mission devoted to West German experiments, mainly materials processing, in pressurized module. Landed 6 November in California.

61B
Atlantis launched 27 November 1985 with Brewster Shaw, Bryan O'Connor; mission specialists Mary Cleave, Sherwood Spring, Jerry Ross; payload specialists Charles Walker, Rudolfo Neri Vela (Mexico). Three comsats are released; Ross and Springer space walk to test space construction methods. Landed 3 December in California.

61C
Columbia launched 12 January 1986 with Robert Gibson, Charles Bolden; mission specialists Franklin Chang-Diaz, Steven Hawley, George Nelson; commercial passenger Robert Cenker and Congressional observer Bill Nelson. Released comsat, carried Material Science Laboratory and 12 GAS canisters. Landed 18 January 1986 in California.

Notes. The early flights carry the 'STS' tags (short for Space Transportation System). NASA changed after STS-9 to a new system. Mission 41G, for example, means it was the 7th planned flight ('G') of financial year 1984 (the '4') and launched from Cape Canaveral (the '1'). Missions from California will have a '2' as the second digit. Also note that NASA's financial year begins in October; thus NASA's 1985 began in October 1984. Some missions (such as 41E) were dropped.

Index

Figures in italics refer to illustrations

A-4 (V-2) missile 9, *9*, 10
A-4b missile 10
Abort-to-Orbit (ATO) 101
ACCESS (Assembly Concept for Construction of Erectable Space Structures) 111, *111*
access arm 42, 43, 53, 56
Acquacade type satellite 94
Al-Saud, Prince Sultan Salman 105, *105*
Allen, Joe 62, 69, 70, *70*, 71, 72, 74, 92
American Airlines 40
Anik C1 comsat 75, 96
Anik C3 comsat 70, 71
Anik D2 comsat 92
Apollo missions 25, 26, 41, 42, 43, 56, 110
Apollo 17 10, 77
Arabsat 1B comsat 105
Ariane rocket 16, 105
Armstrong, Neil 12, 14
ASC-1 comsat 108
ASSET 13
Astro-1 Halley's Comet telescope 115

Atlantis Orbiter (OV-103) 18, 79, 109, 110, 113, 115
Atlas rocket 9, 27
attitude control system 85
Aussat 1 comsat 108
Aussat 2 comsat 110
Auxiliary Power Units 57, 63, 80

B-52 bomber 10, 12, 40, 84
backpacks 25, 51, *83*, 92
Baudry, Patrick 94, 105, *105*
Bell Company 10
Bluford, Guion 77
Bobko, Col. Karol 72, 94, 97, 109, 110
Boeing 747 40, *40*
Boeing Company 13
Bolden, Charles *118*, 119
boost motor 27, 70, 92
Brand, Vance 47, 69, 71, 83
Brandenstein, Dan 59, 77, 105
Bridges, Roy *101*
British Aerospace 66, 78
British Aircraft Corporation 14
Buchli, Lt. Col. James 94, 95
Bush, George 61

cameras 27, 59, 60, *60*, 66, 70, 75, 79, 88, 110
Cape Canaveral 10, 15, 41, 44, *44*, 46, *46*, 47, 57, 76, 85, *85*
Cenker, Robert 115, *118*
Challenger Orbiter (OV-099) 7, 18, 21, 27, 31, 38, 39, 42, 43, 46, *46*, 47, 72, *72*, 74-9 *passim*, 75, 83, 85, 86, 91, *91*, 97, 101, 105, 106, 115, 119, 120, *120*, 121, *121*
CHAMP (Comet Halley Active Monitoring Program) 114
chest-packs 25
Cirrus experiment 68
Cleeve, Mary 112
Coats, Michael 86
colonization of space, human 8
Columbia Orbiter (OV-101) 18, 19, 20, 30, 31, 41, 45, 47, 49-51, *49*, *50*, *51*, 53-4, *53*, *54*, 56-60 *passim*, *60*, 61, 62, *62*, 63, 64, 65, 66, *66*, 67-8, 70, 72, 78, *78*, 79, *79*, 80, 113, 114, 115, *115*, *118*, 119

communications satellites (comsats)
Anik C1 75, 96
Anik C3 70, 71
Anik D2 92
Arabsat 1B 105
ASC-1 108
Aussat 1 108
Aussat 2 110
DSCS-3 military 109, 110
Insat-1B 76-7
Morelos-1 105-6
Morelos-2 110
Palapa B1 74, 75
Palapa B2 83, 92, 94
Satcom K-1 115
Satcom K-2 110
SBS-3 70, 71
Syncom 4-2 88, *88*
Syncom 4-3 74, 96, 97, 106, *106*, 108-9, *108*
Syncom 4-4 108, 109
TDRS 43, 72, 82
TDRS-1 72, 79, 91, 95
TDRS-2 74, 76, 95, 119
TDRS-3 115
Telstar 3 87
Telstar 3D 105-6
Westar 6 74, 83, 92, *92*
computers 21, 43, 49, 53, 56, 57, 58, 62, 63, 75, 80, 85,

125

86-7, 101, 108
Control Data Subsystem 43
controls 20, 21-2
costs 8, 10, 15-16, 29, 36, 70, 78, 85, 96, 98, 105, 106, 109, 119
Countdown Demonstration Tests 54
Covey, Dick 108, 120
Creighton, John 105
crew quarters 20, 21, 22-3, 25
Crippen, Robert 47, 49, 50, 51, 53, 54, 56, 57, 60, 60, 61, 61, 62, 75, 75, 76, 85, 86, 91, 110, 115
Crossfield, Scott 12
crossrange 15, 76
crystals 98, 111, 115

Defense, Department of 67
Delta rocket 85
Discoverer spy-satellite 46
Discovery Orbiter (OV-102) 18, 27, 46, 47, 86, 87, 89, 92, 94, 97, 105, 106, 109, 115
Drop Dynamics Module 98, *98*
DSCS-3 military comsat *109*, 110
Dynasoar project 13

Earth Observing Mission 115
Earth Radiation Budget Satellite 91
EASE (Experimental Assembly of Structures in EVA), 111, 112, *112*
echography 94
Edwards Air Force Base, California 10, *13*, 40, 49, 62, *62*, 68, 72, 75, 76, 77, 106
ejection suits 50, 57, 61, 62
electrophoresis unit 65, 66, 68, 75, *75*, 77, 88, 89, 96, 110
England, Tony 101
Engle, Joe 12, 47, 63, *63*, 108, 109, 110
Enterprise Orbiter 18, *18*, 39, *39*, 40, *40*, 47, 72
Environmental Mobility Unit (EMU) *see* spacesuits
European Space Agency (ESA) 78
exercise 63

Explorer 1 satellite 9
external fuel tanks 7, 16, 19, 20, 33, *33*, 35, 36, 42, *42*, 50, 53, 56, 59, *59*, 67, 72, 101, 120, *120*
Extra-Vehicular Activity (EVA) 45, 57, 74, 92

Fabian, John 75, *75*, 105, 106
Fisher, Anna 23, 92, *106*
Fisher, William 74, *106*, 108, 109
Fixed Service Structure (FSS) 42
flight cabin 20, 21-2
flight simulator *48*, 49
'fly swats' 74, 96, 109
Ford, Gerald 18
Freedom Star (recovery ship) 38
Fullerton, Charles 47
Fullerton, Gordon 39, 40, 64, *65*, 101, 105

Gagarin, Yuri 9, 56
Galileo Jupiter probe 106, 115
Gardner, Dale 74, 77, 92, *92*, 94, 95
Garn, Senator Jake 94, 96, *96*, 110
Garneau, Marc 91, *91*
Garriott, Owen 79, *80*
Gemini missions 25, 56
 Gemini 3 61
 Gemini 4 44
German Society for Space Travel 9
Getaway Special canisters 29, 44, 68, 75, 77, 115
Gibson, Robert 83, 119
Glenn, John 9, 27
Goddard, Dr Robert 8, *9*
Grabe, Col. Ronald 109
Gregory, Fred 99
Griggs, David 74, 94, 96
Grissom, Virgil 49
Grumman aerospace company 98

Haise, Fred 39, 40, 47
Halley's Comet 7, 105, *114*, 115, 119
Harris, Hugh 57-8
Hart, Terry 85, 86
Hartsfield, Henry 67, 86, 87
Hauck, Frederick 75, 92, 110
Hawley, Steven 86, 115

Head Up Display *118*
Heading Alignment Cylinder 62
Henize, Karl 101
Hermes 12
Hilmers, Major David 109
Hitchhiker 115
HL-10 lifting body 13
Hoffman, Jeffrey 74, 94, 96
HS-376 type satellite 70
Hubble Space Telescope 115
Hughes Company 70, 87, 92, 97, 106

Independence (recovery ship) 38
Induced Environment Contamination Monitor (IECM) 66, 68
Inertial Measurement Unit (IMU) 61
Inertial Upper Stage rocket (IUS) 72, *72*, 74, 76, 83, 94, 110, 119
Insat 1B comms/weather satellite 76-7
Instrument Pointing System (IPS) 101, *103*
insurance coverage 92, 110
Intercontinental Ballistic Missiles (ICBMS) 9
Intermediate-Range Ballistic Missiles (IRBMS) 9

Jarvis, Gregory 113, 119
Johnson Space Center, Houston 44, 48, 121
Jumbo jet *18*, 41, 44, 46, 50
Jupiter 8
Jupiter missile 9

Kennedy Space Center 19, 36, 44, 50, 67, 97, 115
KH-11 satellites 60
Ku-band dish antennae 91, 92

Landsat earth-resources satellites 46, 92
laser tracking 106
Launch Control Center 41, 42-3, 44, 54
launch pad 41, 42, 44, 58
 39A 19, *19*, 41, 47, 53, 58, 63, 79, 101
 39B 19
 Fixed Service Structure (FSS) 42

Rotating Service Structure (RSS) 42, *43*, 54
launch schedule (1986) 115
Leestma, David 74, 91, 92, *92*
Lenoir, William 69, 70, 71, 72
Liberty Star (recovery ship) 38
Lichtenberg, Byron 79, *80*
lifting bodies 13, *13*
lightning conductor 19
Lind, Don 99
Long Duration Exposure Facility 85, *85*
Lounge, Mike 108
Lousma, Jack 47, 64, *65*, 67
Lucid, Shannon 105, 106
lunar orbiting base 14
lunar surface base 14

M2 lifting body 13
M2-F2/3 lifting body 13
McAuliffe, Christa 7, 113, 119
McBride, Jon 91
McCandless, Bruce 25, 74, 82, 83, *83*, 84
McDonnell-Douglas Company 110
McNair, Ronald 22, 83, 119
Main Engine Cut-Off (MECO) 59
main engines 7, 19, 20, 27-30, *27*, *28*, 39, 47, 51, 53, *53*, 57, 58, 59, 72, 79, 86, 101, 119, 120, 121
Manned Maneuvring Unit (MMU) 51, 74, 84, *84*, 85
Manned Orbital Laboratory 49
Mars 8, 14
Martin-Marietta Company *33*, 51
Mate/Demate Device 44, 50
Materials Science Laboratory-2 115
Mattingly, Commander Tom 67, *67*, 68, 94, 95
meals 23, *23*, 60, *60*, 61
medical research 65, 77, 88-9, 94, 105, 110, 114, 115
Merbold, Ulf 79
Mercury 8
Mercury spacecraft 9, 10, 56
Milky Way 106
Mission Control Center, Houston 57, 59, 63

Missions Operation Control Room (MOCR) 57
Monodisperse Latex Reactor (MLR) 65, *65*, 66, 67, 68, 75
Moon, The 8, 14, 25, 41, 47
Morelos 1 comsat 105-6
Morelos 2 comsat 110
Morgan, Barbara 119
Morton-Thiokol Company 36
moth/bee flying experiment 65, 66
MSL-2 experiments 115
Mullane, Richard 86
Musgrave, Story 72, 74, *74*
MUSTARD (Multi-Unit Space Transport and Recovery Device) 14

Nagel, Steven 105
NASA (National Aeronautics and Space Administration) 14, 15, 16, 18, 19, 28, 29, 40, 41, 47, 49, 50, 51, 53, 56, 58, 60, 61, 63, 66, 67, 68, 70, 72, 76-9 *passim*, 85, 88, 91, 92, 94, 95, 97, 101, 105, 106, 109, 110, 113, 115, 119, 120
National Advisory Committee for Aeronautics (NACA) 10
National Space Technology Laboratory 51
Nelson, Congressional Observer Bill 113, *114*, 115
Nelson, George 74, 85, 86, *86*, *114*
Nelson, Todd 65, 66
Newton, Isaac 8
Nixon, Richard 16
Nomex Felt Reusable Surface Insulation 31, 50
Northrup Strip, New Mexico 67

Oberpfaffenhofen Centre 101
Oberth, Prof. Hermann 8, *9*
O'Connor, Bryan 113
Office of Space Science (OSS) 66

Onizuka, Major Ellison 94, 95, 119
orbital flight test (OFT) 47
Orbital Maneuvring System (OMS) 27, 29-30, *29*, 43, 54, 60, 62, 67, 92, 101
Orbital Sciences Corporation *115*
Orbiter Processing Facility (OPF) 43, 44, 45, 50, 53, *78*
Orbiters 20-23, 25-31 (main general references) *see also* Atlantis; Challenger; Columbia; Discovery
Overmyer, Bob 69, 70, *71*, *99*, 101

Page, George 54
Pailes, Major William 110
Palapa B1 comsat 74, 75
Palapa B2 comsat 83, 92, 94
Parker, Robert 79, *80*
Payload Assist Module boost motor (PAM) 70, 77, 83, 87, 92
payload bay 15, 21, 22, 25, 26-7, 44, 66, 68, 70, 72, 76, 78, 110
payload bay doors 26-7, 31, 60, 61
Payload Flight Test Article (PFTA) 76, 77
Payton, Major Gary 94, *95*
Pentagon 94, 121
Peterson, Don 72, 74, *74*
Plasma Diagnostics Package (PDP), 66, *66*, 67
pollution 8, 66-7
Portable Life Support System *see* backpacks
Presidential Space Task Group 14
PRIME 13

radar experiments 64, 65
radio blackout 62
RCA Company 110, 113, *118*, 119
Reagan, Nancy 69
Reagan, Ronald 68, 69, 94, 121
recovery ships 38, *38*
Redstone missile 9, *9*
refuelling techniques 92, *92*
Remote Manipulator System *see* Robot arm

Resnik, Judith 86, 89, 119
Return to Launch Site (RTLS) abort 58, 59
revenue 8, 105, 110
Ride, Dr Sally 63, 75, *75*, 86, 91-2, 110
Robot arm (Remote Manipulator System) 22, 25, 27, *27*, 29, 60, 63-4, 66, 68, 69, 74, 76, 77, 82, 86, 87, 88, 89, 91, 92, 96, 97, *97*, 108, 112, 119
rocket thrusters 21, 27, 30, 60, 61
Rockwell Corporation 16, 17, 18, 39, 50, 56, 86, 109, 113
Ross, Jerry 74, 110, 111, 112, *112*
Rota US Naval Air Station 59
Rotating Service Structure (RSS) 42, 43, 54
runways 44, *46*, 62, 67, 76, 85, 97

Salyut space-station 23
Satcom K-1 comsat 115
Satcom K-2 comsat 110, 113, 115
satellites
  early warning 68
  earth-resources 8, 46, 92
  spy 60, 94
  weather 8, 46, 76-7
  see also communications satellites; individual names of satellites
Saturn 8
Saturn rockets 9, 19, 29, 42, 43
Saturn 1B 10
Saturn 5 10, *10*, 41, 42
SBS-3 comsat 70, 71
SBS-4 comsat 87
Scobee, Francis 85, 119, 120
Scout rocket 10
Scully-Power, Paul 91, *91*
Seddon, Rhea 23, 94, 97, 109
Shaw, Brewster 79, 80, *113*
Shriver, Lt. Col. Loren 94, 95
Shuttle Carrier Aircraft (SCA) 40
Shuttle Imaging Radar 91
Shuttle trainer 22, *23*
silica shield 30-31
Skylab 10, 22, 23, 61, 110
sleeping facilities 23, *23*, 25, 61

Smith, Michael 119
Smith, Richard 44
Solar Maximum Mission satellite (Solar Max) 74, *82*, *83*, 85, 86, *86*, 87, 92
solar panel 77, 85, 86
solid rocket boosters (SRB) 14-15, *14*, 16, *16*, 19-20, 27, 29, 36, 37, 38-9, *38*, 42, *42*, 50, 58, *58*, 59, 101, 119, 120, *120*, 121, *121*
space engineering 113
Space Shuttle Main Engines (SSME) *see* main engines
Space Shuttle Missions
  STS-1 47, 50, 54, *54*, 56-62, *58*, *59*, 60, 61, 62, 63, 124
  STS-2 38, 45, 47, 62, 63-4, 91, 124
  STS-3 30, 47, 64-7, *65*, 124
  STS-4 44, 47, 64, 65, 67-8, *67*, 69, 124
  STS-5 26, 38, 68-9, 70, *70*, 71, 72, 124
  STS-6 42, 43, 72, *72*, 74-5, *74*, 95, 124
  STS-7 27, 29, 75-6, *75*, 124
  STS-8 42, 76-7, *76*, 77, 79, 124
  STS-9 78-80, *78*, *79*, *80*, *81*, 124
  STS-10 94
  STS-11 77
  41B 25, 74, 82, *83*, 83-5, *83*, *84*, 92, 124
  41C 74, *83*, 85-6, *85*, *86*, 87, 124
  41D 22, 87-9, *88*, *89*, 119, 124
  41G 74, 91-2, *91*, *92*, 124
  51A 23, 74, 92, *92*, 94, 124
  51B 98, *99*, 101, 124
  51C 94, 95, 119, 124
  51D 74, 85, 94, 95-7, *95*, 106, 124
  51E 94-5
  51f 101, *101*, 105, 125
  51G 105-6, *105*, 125
  51I 74, 97, 106, *107*, 108-9, 113, 125
  51J 109-10, *109*, 125
  51L 7, 74, 94, 95, 101, 105, 115, 119-21, *120*, *121*
  61A 23, 101, 125
  61B 74, 110-13, *111*, *112*, 125
  61C 113, *113*, 115, *118*,

119, 125
61E 105, 115
61G 106, 115
space sickness 67, 72, 75, 77, 96, *96*
space spheres 65
space stations 8, 14, 41, 44, 68, 110, 111, 115
space walks 21, 25, 68, 70, 72, 74, *74*, 84, 85, 96, 111-12, *111*
Spacelab modules 19, 44, 97
Spacelab 1 78, *78*, 80, 98, 101
Spacelab 2 79, 98, 101, *101*, 105
Spacelab 3 98, *99*, 101, 105
Spacelab D-1/61A flight 101, 105
spacesuits 25-6, *25*, 56, 57, 68, 69, 72
Spartan retrievable satellite 106, 119
SPAS-01 satellite 75, *75*
Spring, Sherwood 74, 110, 111, *111*, 112, *112*
Sputnik 1 satellite 9
Star Trek 18
Star Wars 105, 106, 115
Stewart, Col. Robert 74, 83, 84, 109
Structural Test Article-099 18
Sullivan, Kathryn 74, 91
Sun, The 66, 80, 85, 105
sunshield 70, *71*, 108

Syncom 4-2 comsat 88, *88*
Syncom 4-3 comsat 74, 96, 97, 106, *106*, 108-9, *108*
Syncom 4-4 comsat 108, 109, 113

TDRS (Tracking and Data Relay Satellite) 43, 72, *72*, 82
TDRS-1 72, 79, 91, 95
TDRS-2 74, 76, 95, 115, 119
TDRS-3 115
Teacher in Space Project 119
Teal Ruby satellite 115
telemetry analysis 121
telescope 85
  Astro-1 Halley's Comet 115
  astronomical 66-7
  Hubble Space 115
  infrared 68, 105
  solar *86*
  ultra-violet 80
  X-ray 80, 105, 106
Telstar 3 comsat 87
Telstar 3D comstat 105-6
Thagard, Dr Norman 75, *75*, 99
thermal protection 13, 15, 20, 21, 25, 27, 30-31, *30*, *31*, 33, 36, 43-4, 47, 50-51, *51*, 54, 60, *60*, 62, 63, 67, 68, 70, 72, 86, 109
Thor missile 9

Thornton, Dr Bill 77, *77*, 99
3M Corporation 110-11
Tiros satellite 46
Titan missile 9
Titan 1 missile 13
Titan 3 missile 121
toilet facilities 22
Tracking and Data Relay Satellite *see* TDRS
Truly, Richard 47, 63, 77
Tsiolkovsky, Konstantin 8, 121
turbopumps 29

Ulysses Sun probe 115
Uranus 7, 8
U.S. Air Force 10, 12, 13, 46, 49, 57, 60, 67, 72, 94, 110, 115
U.S. Navy 10, 87, 106, 108
Utah, University of 68

V-2 missile (A-4) 9, *9*, 10
Van den Berg, Lodewijk 99
Van Hoften, James 'Ox' 74, 85, 86, 87, 108-9, *108*
Vandenberg Air Force Base Launch Site 15, 41, 46, 47, 115
VAB (Vehicle Assembly Building) 19, *19*, 41, *41*, 42, *42*, 44, 50, 53, 54, 79
Vela, Rudolfo Neri 110
Venus 8

'Vomit Comet' aircraft 119
Von Braun, Wernher 9, *9*, 10
Vostok 1 spacecraft 9
Voyager 2 space probe 7

Walker, Charles 65, 86, 88, *89*, 95-6, 110
Walker, David 92
Wang, Taylor 98, *98*, 99
water system 21
weightlessness 12, 23, 66, 68, 98, 105, 115, 119
Weitz, Paul 72, 75
Westar 6 comsat 74, 83, 92, *92*
Williams, Don 94
wind-tunnel tests 17, 40

X-1 aircraft 10
X-15 aircraft 10, 12, *12*, 13, 63
X-20 Dynasoar shuttle craft *12*, 13
X-24A lifting body 13
X-24B lifting body 13, *13*

Yeager, Charles 10
Young, John 47, *48*, 49, 50, 53, 54, 56, 57, 60-61, *60*, *62*, 79, 80

zero-gravity 66, 115